PROCEEDINGS OF THE 30TH INTERNATIONAL GEOLOGICAL CONGRESS
VOLUME 13

MARINE GEOLOGY AND PALAEOCEANOGRAPHY

Proceedings of the 30th International Geological Congress

PROCEEDINGS OF THE
30TH INTERNATIONAL
GEOLOGICAL CONGRESS

BEIJING, CHINA, 4 - 14 AUGUST 1996

VOLUME 13

MARINE GEOLOGY AND
PALAEOCEANOGRAPHY

EDITORS:
WANG PINXIAN
DEPARTMENT OF MARINE GEOLOGY, TONGJI UNIVERSITY, SHANGHAI, CHINA
W.A. BERGGREN
WOODS HOLE OCEANOGRAPHIC INSTITUTION, WOODS HOLE, MA, USA

CRC Press
Taylor & Francis Group
Boca Raton London New York

CRC Press is an imprint of the
Taylor & Francis Group, an **informa** business

First published 1997 by VSP BV Publishing

Published 2019 by CRC Press
Taylor & Francis Group
6000 Broken Sound Parkway NW, Suite 300
Boca Raton, FL 33487-2742

© 1997 by Taylor & Francis Group, LLC
CRC Press is an imprint of Taylor & Francis Group, an Informa business

First issued in paperback 2019

No claim to original U.S. Government works

ISBN 13: 978-0-367-44817-2 (pbk)
ISBN 13: 978-90-6764-242-2 (hbk)

Visit the Taylor & Francis Web site at
http://www.taylorandfrancis.com

and the CRC Press Web site at
http://www.crcpress.com

CONTENTS

Proc. 30ᵗʰ Int'l. Geol. Congr., Vol. 13, pp. 1-32
Wang & Berggren (Eds)
©VSP 1997

Geophysical Survey in the South China Sea with a Special Focus on the Backarc Basin Formation

HAJIMU KINOSHITA
Japan Marine Science and Technology Center, 2-15 Natsushima, Yokosuka, Japan
YUKARI NAKASA
Ocean Research Institute, University of Tokyo, 1-15-1 Minamidai, Nakano, Tokyo, Japan
and
LIU ZAOSHU and XIA KANYUAN
South China Sea Institute of Oceanology, Academia Sinica, 164 West Xingang Road, Guangzhou, China

Abstract

Basic data on the opening of backarc basin, South China Sea (abbreviated as SCS hereafter) basin, were obtained through a two-year Japan and China cooperative joint marine geophysical survey in 1993 and 1994. Some new gravity, magnetic and seismic structural data by reflection as well as refraction study of SCS were obtained and added to existing data. A systematic three component magnetic measurement was made for the first time in this basin. These data in addition to those of earlier studies are constraints for depicting a thinning of a continental crust and later followed by a backarc basin opening by ocean floor spreading. Gravimetric and magnetic anomalies, heat flow, and ages of drilled rocks from this area are referred to speculating on the opening scheme of SCS. Origin of disoriented and scattered magnetic anomaly lineations is suggested to be due to intermittent activities along opening ridge systems of SCS.

Keywords: backarc basin, seismic structure, magnetic anomaly, gravity anomaly, evolution

INTRODUCTION

SCS is located between the continental and oceanic tectonic domains in the midst of intersection of Eurasian, Pacific and Indian plates (Fig. 1). The basin has been the

----

Note: List of participants in the Japan-China joint study for 1993-1994 is given in the appendix.

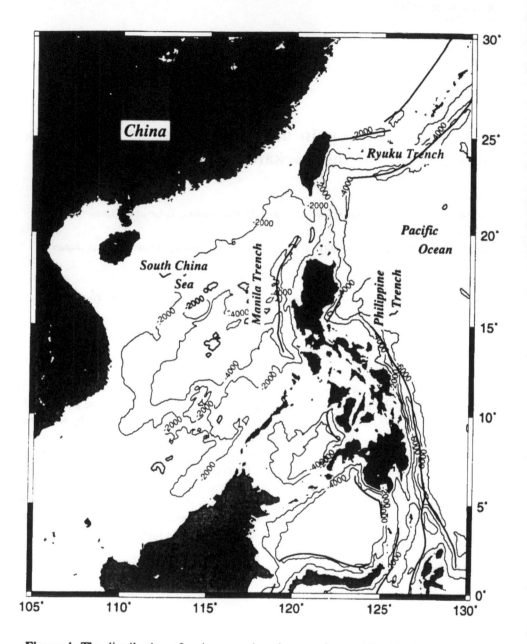

Figure 1. The distribution of major tectonic units, trenches and Pacific Ocean around the South China Sea (SCS).

objective of extensive geoscientific studies. One of the benefits of studying characteristics of SCS is that it provides an ideal site for the study of passive continental margins. It is likely that SCS is old enough to have experienced complete history of a passive margin development, yet it is young enough to have the records of its history preserved. This paper discusses a possible mechanism for the crustal extension in the northern SCS based on number of geophysical parameters and, in addition, on newly obtained orientation of the spreading as well as the deep crustal structure. One of the purposes of our present study is to define in detail the segmented spreading axes by estimating the orientations of magnetized bodies which produce magnetic lineation patterns in the oceanic basin of SCS.

SCS is important also in understanding the tectonic evolution of the western Pacific because SCS is a major constituent among many marginal seas that characterize the western Pacific. Furthermore, the global significance of initiation of the marginal sea opening may come under light by comparing SCS evolution with rifting and break up of continent-ocean boundaries along the north Atlantic margin, one of the world best studied areas.

The northeastern boundary of SCS is surrounded by Taiwan island arc which is obviously controlled by the NNE* tectonic structure; the eastern boundary is Philippine island arc which is controlled by the NE to NNW tectonic structures; the southern boundary is Kalimantan island arc; the western boundary is apparently controlled by the offshore fault of Vietnam. These tectonic units were originated in the Himalayan period accompanied with the formation of Hereynian relic block and Yenshanian basement.

Based on the recent geological and geophysical data (Ke, 1985), there is a marginal suture line composed of Cretaceous and early Eocene volcanic formations in the S-N* direction from the southeastern offshore of Indo-China to the south of west Kalimantan. The suture divides Sunda shelf from Nansha depression.

The horizontal stress distribution within the crust consists of apparent tension in the northern margin and compression in the southern margin (Xia and Zhou, 1993). The thickness of the crust is heterogeneous. Orientation of apparent basin ridge systems, timing of opening, and rate of deformation of the crust show complicated features reflecting the heterogeneity of tectonic structures. A double tectonic zone syngenetic of different nature was formed in Yanshanian where the continental margin was transformed into the oceanic basin (Xia et al., 1994). The thickness of the oceanic crust from the seismic velocity structure is significantly different from that of the western Pacific (Xia and Zhou, 1993).

There is a variety of speculative interpretations on the formation of SCS. These speculations are mainly based upon geometrical considerations. Taylor and Hayes (1980-b) suggested that SCS had evolved with a change in the orientation of opening from N-S to NW-SE in the eastern part of SCS. Ru and Pigott (1986) suggested that there were two stages of spreading and three rifting episodes. Briais et al. (1989) noted that there is some ambiguity in the change of spreading orientation and suggested that there can be number of alternative interpretations based on the same magnetic lineation data. Lee and Lawver (1992) mentioned that the whole opening history of SCS could be clarified by examining structural fabrics, paleostress patterns, geological data, and paleomagnetic data in addition

---

*Orientation of geological units is abbreviated as NNE which reads north-north-east throughout this report. Also S-N denotes north to south stretch.

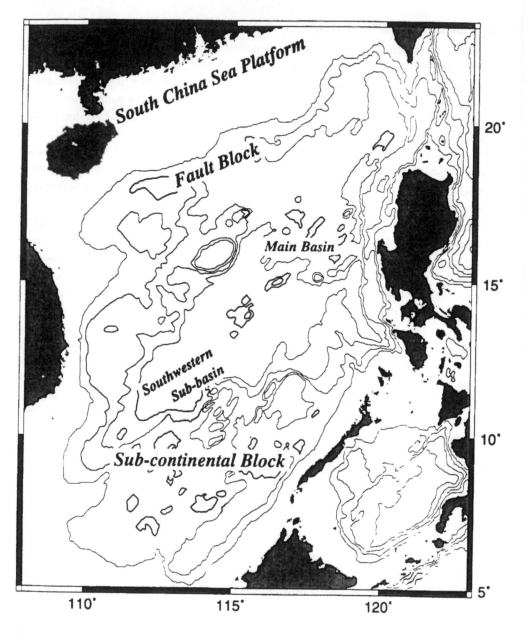

Figure 2. The South China Sea is classified into number of tectonic units; South China Sea platform, South China Sea fault block, main basin, southwestern sub-basin and sub-continental block in its southern part.

to the regional geological setting of the past.

The SCS basin is classified into a number of tectonic units (Fig. 2). The age of the north of the main basin is estimated from heat flow data as 45 Ma to 51.1 Ma, whereas the southwestern sub-basin is approximately 15 Ma (e.g., Parsons and Sclater, 1977; Taylor and Hayes, 1983; Matsubayashi and Nagao, 1991).

It seems from the age distribution in SCS that the NE aligned fault system was activated after the spreading of the Central basin associated with extension of the continental margin (Li 1973; Huang and Ren, 1980; Liu, 1984; Liu et. al., 1984; Liu and Chen, 1987; Taylor and Hayes, 1983; Chen, 1985). The spreading activity of SCS propagated westward to form the southwestern sub-basin. The oceanic crust of the Central basin is surrounded by fault systems associated with a large gradient in magnetic as well as gravity values and sharp change in the crustal thickness.

The central part of the basin has many topographic highs, in particular the Scarborough seamount chain, which indicates recent volcanic activity of ca. 12 Ma (Hekinian et al., 1989). The basin can be thus divided morphologically into eastern and western parts at the boundary of 115°E longitudinal line. Hereafter, the eastern part is named main basin and the southwestern part is southwestern sub-basin. In the northwestern part of the basin, the Yingge-hai basin, between Hainan and the coast of Vietnam, is a large sedimentary basin

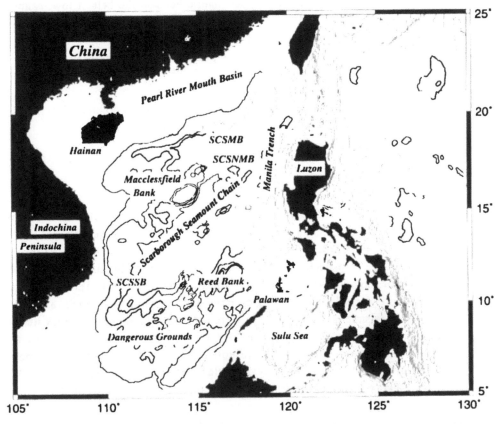

Figure 3. The distribution of geologic units and abyssal basins in the South China Sea. SCSMB, SCSNMB and SCSSB denotes SCS marginal basin, northern part of SCS basin (equivqalent to main basin used in Fig. 2 and in the text) and southern part of SCS basin (equivqalent to southwestern sub-basin used in Fig. 2 and in the text), respectively.

(Letouzey et al., 1988).

The main basin has been subducting at the Manila trench (Pautot et al., 1986; Rangin et al., 1990-a and -b), which elongates from north to south. On the other hand, the southern part of the main basin, Nansha trough and Reed bank, and Macclesfield bank consists of micro-continental blocks, probably a remnant of the southern margin of China (Hinz and Schulter, 1985; Lee and Lawver, 1994) (Fig. 3).

The gravitational compensation mechanism of SCS is studied using admittance functions (e.g., Louden, 1981; Verhoef and Jackson, 1991). From the admittance response of SCS, two different compensation models must be considered, i.e., Airy-type and an elastic plate type. The results of the present study indicate that the local isostasy held in most part of the SCS basin.

There is a 200 km long magnetic and gravity low zone striking E-W and ca. 100 km wide along the continent and oceanic transition zone (COB hereafter; Fig. 4). A seismic high velocity layer with Vp (longitudinal seismic wave velocity) 7.0 to 7.3 km/s is identified between lower crust and the upper mantle beneath the belt of low magnetic anomaly zone (Sekine et al., 1994). It implies that mafic mantle materials were underplated to the stretched continental crust.

A special attention is focused on the following problems in this report;
(1) Reconstruction of the opening history of SCS.
(2) Relationship between topography and gravity.
(3) Forward modeling of reconstruction of magnetic and gravity low around COB.
(4) Rifting of SCS in comparison with that of the north Atlantic margins.

SURVEYS

A geophysical survey was conducted, in the northeastern sector of SCS by the Japan-China joint study. The objective of this survey was to clarify the variation of crustal structure in and around COB. Fifteen ocean bottom seismometers (OBS) were used to obtain seismic signals from controlled sound sources. Three profiling lines were run: one NWN-SES line of about 390 km and two E-W lines about 160 km (Fig. 5). Signals from the airguns were recorded by a single-channel hydrophone streamer and OBS simultaneously. Seismic reflection surveys were conducted on number of track lines to delineate seismic formations of the topmost part of the subbottom layers. Gravity and three component magnetic measurements (Shipboard Three Component Magnetometer: STCM) were also run along track lines. Wide angle reflections were also recorded by OBS which provide a good constraint on the seismic structure with high heterogeneity.

SEISMIC AND GEOLOGICAL STRUCTURES

The sedimentary cap is mainly composed of Cenozoic over the entire SCS area, and locally underlain by relict Mesozoic. The basement rock and the crustal structure show a large inhomogeneity. In terms of sediment distribution, SCS can be divided into three parts from north to south; the northern epicontinental block (South China continental block), the oceanic basin (SCS basin) and the southern epicontinental block (Nansha continental block) (South China Sea Institute of Oceanology, 1986; Xia et al., 1994).

The north epicontinental block includes the Pearl river mouth basin and an uplifted margin to its south. Its northern margin is composed of half-graben and stepped faults controlled by southward dipping normal faults. The central and the southern parts are occupied by plenty of half-grabens and faulted domes controlled by northward dipping faults.

The Pearl river mouth and Beibu basins consist of Paleogene large-scale extensional features, parallel to the shoreline (Taylor and Hayes, 1983). The half-graben is filled with upper Paleocene to lower Oligocene strata, mainly fluviatile sediments (Letouzey et al.,1988).

The oceanic basin (main and southwestern sub-basin) reaches water depth 3 to 4 km.

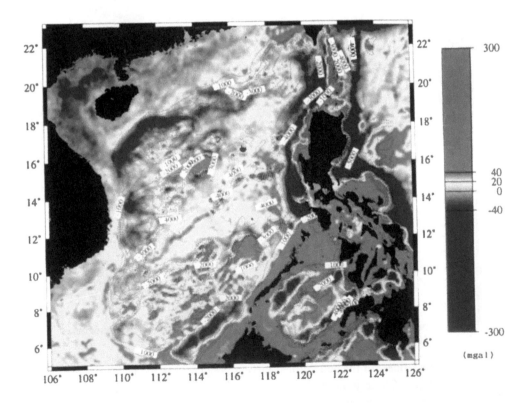

Figure 4. Free-air gravity anomaly contour map constructed by GMT dataset combined with number of recent surveys. The anomaly trend strikes parallel to the coast line ranging from -20 to +20 mgal. Deep negative along N-E coast line to the south of Hainan island is referred to as part of COB in the text.

Both its northern and southern borders are limited by normal faults dipping inward. This area can be subdivided as central basin, Zhongsha north ridge and northwest basin. Sea floor spreading produced two basins (main basin and southwestern sub-basin) with rugged sea floor and domed central ridges, a trace of spreading axes.

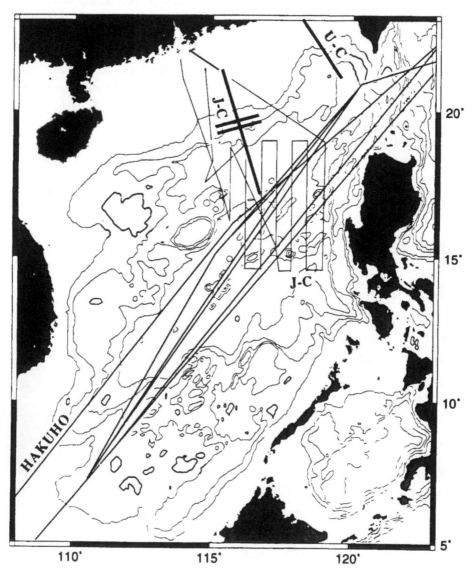

Figure 5. Track lines for the geophysical surveys conducted in the northeastern sector of SCS by the Japan-China joint study in 1993-1994 (J-C). Seismic reflection surveys, gravity and three component magnetic measurements were conducted along these lines. Thin lines (HAKUHO attached) are track lines of R/V Hakuho-maru, University of Tokyo. U-C denotes one of track lines by USA-China efforts (further details are given in the text).

The south epicontinental block includes Liyue bank (Reed bank) and Nansha trough region and is of extensional structure. Its northern part is lifted while southern part is subsided, and the inner part is considerably extended to form a series of half-graben and faulted domes.

Macclesfield, Paracel, Reed bank, North Palawan, and Dangerous grounds (Nansha trough) are remnants of continental blocks (Letouzey et al., 1988). According to seismic reflection profiles, the structures is similar both in the northwest and southeast margins of SCS (Lee and Lawver, 1994; Xia et al., 1994).

Seismic structure of upper crust from former studies

The basement layer is clearly separated from the sedimentary cap. Three structural supersequences of sedimentary layers can be identified above the basement (Chen, 1987). Explanation for crustal structures is given in the Atlas of Geology and Geophysics of South China Sea (Chen, 1987). The basement layers are divided by two regional unconformities which represent two major tectonic events during Mesozoic and Cenozoic, respectively. The upper layer overlies Oligocene formation and consists of horizontally varying components over the entire region including upper Eocene to Recent in the northern margin.

Middle layer consists mainly of Paleocene to Eocene strata. Its age starts from Late Cretaceous epoch and ends at middle Eocene. This layer develops widely over the southern margin. This middle layer, however, is not found in the basin. Therefore, it is likely that the SCS basin did not exist before Oligocene.

Upper basement layer is identified mainly in Liyue beach (Reed bank) and Nansha trough in the southern continental margin. Drill and dredge-haul samples show that there are upper Jurassic-lower Cretaceous shale, upper Triassic-lower Jurassic sand-mudstone, and middle Triassic abyssal shale.

The rock beneath the sonic basement is igneous-metamorphic. The basement varies in age and formation. In the abyssal basin area, the basement consists of basalt, product of seafloor spreading during early Oligocene-early Miocene (32-17 Ma) as estimated from magnetic anomaly models. Some seamounts are products of lava eruptions since Miocene after cessation of seafloor spreading. The northern marginal region has a basement of intermediate-acidic igneous, provided by twenty seven drill holes. The K-Ar isotopic ages of rocks from this region are 70.5 to 130 Ma, i,.e., Cretaceous (Xia and Zhou, 1993).

Two-ship seismic expanded spread profiles and sonobuoy dataset revealed that the crust and lithosphere of SCS is thinning step by step toward oceanic basin (Hayes et al., 1995; Nissen et al., 1995). These data reflect tectonic crustal thinning of the SCS marginal regions by tectonic extension.

Oceanic basement from former studies

The reflective basement topography has no correlation to the sea-floor topography. Ocean floor is fairly smooth from the continental shelf to the basin, compared to a fairly rugged acoustic basement. According to three profiles across COB (Hayes, 1992), no seismic high velocity layer could be defined or does not exist under the eastern COB in contrast to the present results.

The oceanic basement can be divided into two layers with different velocities, corresponding to the oceanic layer 2 and layer 3, respectively (Hayes, 1988, 1992; Chen, 1995). The layer 2 has velocity 4.6 - 6.2 km/s and thickness 1.5 - 3.5 km. The layer 3 is 6.6 - 7.4 km/s in velocity and 1.5 - 3 km in its thickness.

In the northern margin, the basement has upper and lower crustal layers, where the upper one has velocity 5.2 - 6.1 km/s and thickness only 1 - 3 km. The lower layer has velocity 6.5 - 7.0 km/s with their thickness undefined.

In the southern margin area, velocity data from refraction study are not available so far. According to velocity analyses of reflection seismic data, the interval velocity of the upper basement varies from 3.5 to 5.9 km/s.

Seismic structure obtained by present study

Seismic refraction and reflection surveys were conducted over the northern margin of SCS by the present joint study. A two-dimensional ray tracing method was applied to the seismic record sections. The entire crustal structure in the northern part of SCS to the depth of upper mantle could be clarified for the first time (Fig. 6; Sekine et al.,1994). This result indicates that the crustal layer thins gradually toward the oceanic basin, to half the thickness of the continental shelf. The transition zone seems to extend about 100 km. Underneath the lower crust there is obviously a seismic layer with velocity values significantly higher than those of the surrounding layers of the same horizontal levels. It is likely that the continental crust was stretched, extended more than 200 km in the ocean-continent transition zone around present study area.

Summary of geological structures

Summarizing all the information available so far, the geological structure of the SCS basin is described as follows.

1. In the northern and southern margin areas, the cap layers are mainly of Mesozoic and Cenozoic. Unconformities and structural layers show very similar stratigraphy in the north and in the south.

2. The abyssal basin has oceanic basement with only upper Oligocene-Recent deposits over it. Therefore, it is most likely that the SCS abyssal basin was formed after early Oligocene.

3. Seismic high velocity material is found only in a restricted part beneath the continent to ocean transition zone.

4. Upper Cretaceous-Eocene and upper Eocene-Recent have double layer structure, with the lower one half-graben filled and upper one onlapped.

MAGNETIC DATA

There is a clear magnetic low accompanied by a gravity low (Xia et al., 1994) in COB. Taylor and Hayes (1983) suggested that the magnetic quiet zone (MQZ, hereafter) is one of the reflections of COB. The survey area by the Japan-China joint program covers the northeastern part of SCS. The track lines run over COB. The eight N-S parallel ship tracks with its lateral spacing ca. 50 km cover anomaly Chron 5d to Chron 11.

Data from former studies

The marine geomagnetic anomaly data are taken from those open to public, i.e., National Geophysical Data Center (NGDC) of the National Oceanic and Atmospheric Administration (NOAA), U.S.A., total and three-component magnetic data obtained by Japan-China cooperative expedition 1993-1994 and by Hakuho-maru (Ocean Research Institute, Japan) 1989 and 1993 cruises.

The magnetic anomaly profiles of NGDC data were adjusted to IGRF90 (International Geomagnetic Reference Field 1990, IAGA Division V Working Group 8, 1991). Transformation to grid data at 10 km mesh of magnetic field was performed using Akima's method (Akima, 1970 and 1974). The magnetic anomaly contour maps were drawn by applying Briggs' method (Briggs, 1974).

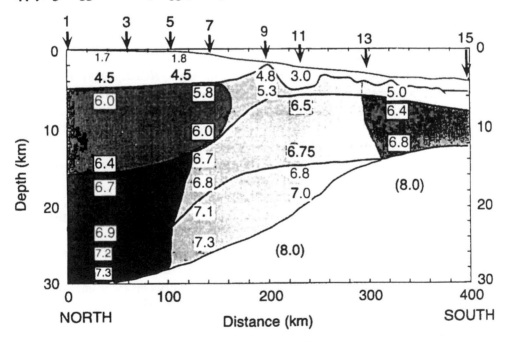

Figure 6. Result of seismic refraction study from fifteen seimic stations where a two-dimensional ray tracing method was applied to the seismic record sections. Depth, distance from shore line (refer to Fig. 5) are given to vertical and horizontal axes. Seismic P-wave speed values are given in units of km/s. Of fifteen stations, location of 8 stations are located on the figure.

Three component magnetic field

An STCM system was used for the present survey. Measurements of total force and three components of the geomagnetic field are designed to obtain more quantitative information of the magnetic field compared to proton magnetometer solely. The magnetic field are assumed to be produced by a linear superposition of the local magnetic field and ship-induced magnetic field. Advantage of STCM is that we can determine the orientation of the elongation of a magnetized body by running over it only once on an assumption that its elongation is large enough compared to the distance between the sensor and the center of magnetic source body (Isezaki, 1986).

A drawback of STCM is a fairly large change in its sensitivity with its temperature due to high-mu metal alloy sensor. A proton magnetometer for the total force was towed behind the vessel to cancel out the variation of the sensitivity of STCM by data processing.

Determination of the direction of magnetic anomaly patterns was achieved by a pseudo-stacking method (Nakasa and Kinoshita, 1994). STCM data provide a quantitative information on magnetic lineament orientation (Isezaki, 1986; Korenaga, 1995). These data afford also a position and a strike of the magnetic polarity boundaries (e.g., Seama et al. , 1993).

Data processing

The study area was divided into number of sub-areas because of different magnetic data density. Magnetic anomalies are compared with topography from ETOPO5 map (5' by 5' grid data of the worldwide bathymetry/topography, NOAA, NGDC, 1988). The anomaly patterns are compared to synthetic profiles calculated by two-dimensional block model (Labreque, 1986) based on the "Geologic Time Scale" (Harland et al., 1989). Synthetic profiles are composed by changing layer thickness, magnetization intensity and skewness. Combination of these parameters are fixed as good enough by iteration.

The results of total and three component magnetic STCM survey in the north-eastern sector of SCS are summarized as follows (Fig. 7).

(1) A conspicuous magnetic quiet zone is defined in the northern fringe of SCS.

(2) Clear magnetic anomaly lineation trend as old as 32 Ma (Chron 11) is identified in the northern end of the abyssal basin.

(3) Continuous sea-floor spreading (Chron 11 to 5) with half spreading rates ranging from 22 to 33 mm/yr occurred over the whole northern abyssal basin.

There seems to be a number of faults or fractures which offset magnetic lineation patters, especially with a conspicuous fragmentation around Scarborough seamounts in the central part. The orientation of the lineation pattern varies from place to place due to segmentation of the ridge system. Trend of lineation patterns were determined to be N 50°E with 10 degree ambiguity. It is probable that the opening scheme of the basin during the active period was not homogeneous not only through time but also from segment segment of the ridge system.

The orientation of the ocean floor spreading is obtained by use of STCM data. Combination of the present data with the data obtained by Hakuho-maru cruise of 1988 and 1993 (Fig. 5), it is shown that the orientation of the seafloor spreading occurred first in N-S direction and later its orientation changed to NNW-SSE after 20 Ma. This change in the general orientation of the seafloor spreading of SCS is in accordance with previous studies (e.g. Briais et al., 1993).

MAGNETIC QUIET ZONE IN COB

The magnetic field anomaly of SCS is two-fold: one is a conspicuous MQZ in COB; and the other is a series of lineation patterns in the central basin area as described in the previous section. Blow up of the MQZ shows that this anomaly extends along the margin with a large sinistral offset in the area to the south of Pearl river mouth. There are two possible explanations for MQZ. One potential cause is a unipolar and negative magnetization of the formation during the Cretaceous long lasting reversal epoch. The other possibility is a result of remagnetization or demagnetization of a restricted region within the crust due to some tectonic activity such as vigorous hydrothermal bleaching.

The depth of the top of the fourth layer is 12-25 km which is most probably deeper than

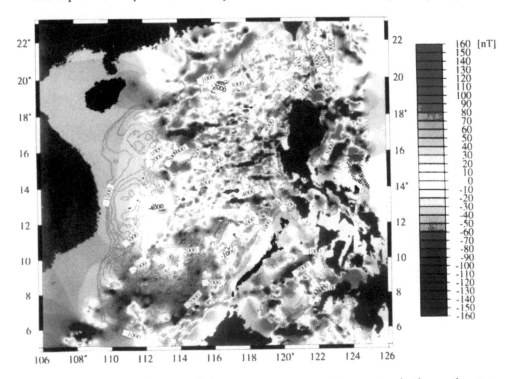

Figure 7-1. Results of total and three component magnetic surveys in the northeastern sector of the South China Sea. A conspicuous magnetic quiet zone exists in the northwestern corner of the main basin (refer also to Fig. 4). Seismic data suggest that it is underlain by a high velocity formation over the mantle. A clear magnetic anomaly lineation trend appears in both of the main and southwestern basins (refer to Fig. 2).

Curie surface (depth at which magnetization disappear due to heating). Thus it is assumed that there are only two potential layers which carry magnetization; the upper crust below sediment cover and the upper part of the lower crust.

It is noted that he intensity of magnetization of materials for this layer in MQ7 ranges from 2.5 to 4.5 A/m with inverse polarizations (Zhang; personal com.). This value is considerably low compared to natural remanent magnetization (varying around 10-50 A/m) of typical midoceanic basalts which seem to resemble basement rocks of the SCS abyssal basin. More detailed description of forward modeling will be given in section 8 of this paper.

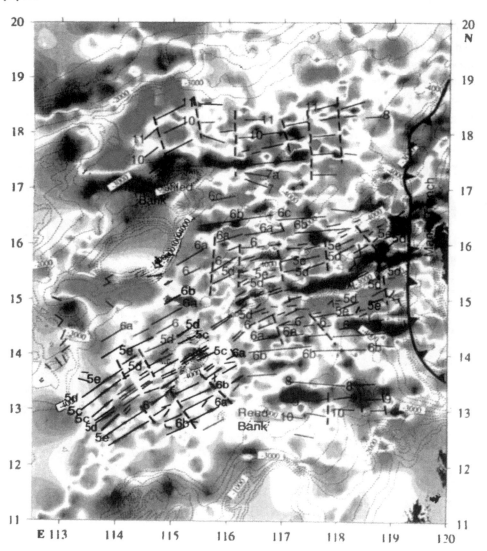

Figure 7-2. The magnetic lineation patterns in the main basin and southwestern sub-basin are identified and given their paleomagnetic age by model calculation.

GRAVITY

The free-air gravity anomaly (FA) contour map is drawn on a background of bathymetry lines (e.g., Sandwell and Smith, 1992) (Fig. 4). The low amplitude FA ranging from -20 to +20 mgal appears parallel to the coast line. The gravity low or negative value can be produced by sedimentary depressions and the gravity high may correspond to the rifting area. The amplitude of FA is mostly in range of -20 to +20 mgal in the whole SCS basin. A higher anomaly values +120 mgal are located in the central mountain area of Taiwan and the minimum -130 mgal is located in the south of Manila trench. FA in the SCS main basin strikes to NE or ENE in general. The amplitude ranges from -25 to +5 mgal in the southern shelf and the relative high is corresponding to the island area.

A linear low gravity anomaly from 0 to -20 mgal trending from NE-SW is a trace of spreading axis. A similar feature in the Coral sea basin has been explained to be due to the mass deficit of the crust and mantle interface (Wessel and Watts, 1979). High gravity anomalies from +340 to +30 mgal are observed along the spreading axis in the main basin; Scarborough seamount chain (5 Ma: Hekinian et al., 1989). Taylor and Hayes (1980-a and -b, 1983) suggested that the landward boundary of oceanic crust is marked by FA low. Relatively low amplitude of FA in entire SCS region suggest that the isostatically compensated mechanism operates throughout the abyssal basin area at large (Louden, 1981). The high gradient zone in the margin of the basin (ca 10 mgal/km) implies that the continent-basin boundary is framed by a confrontation of drastically different materials with a density contrast suggesting a presence of faults into the upper mantle.

The Bouguer gravity anomaly (BA) in SCS increases from continental shelf (-10 mgal) to the main basin (+340 mgal). The anomaly trends towards approximately E-W in the main basin, N-S in the eastern and western margin and NE-SW or ENE-WSW in the northern and southern margin of SCS forms a rhombic configuration with a long axis in the direction of NE.

MECHANICAL AND GRAVITATIONAL STABILITY

Gravitational compensation of the crust over the Moho can be investigated by using an admittance function between topography and gravity (Airy, 1855; Watts, 1978). Three models are usually tested, i.e., 1) Complete isostasy; 2) Elastic and 3) Dynamic balance. The third case has to be applied where there is enough amount of time dependent regional data. We can apply former two methods to better understand the compensation mechanism of SCS.

Gravity data obtained in 1993 and world gravity data (Sandwell and Smith, 1992) are used. Satellite gravity and topography data across the continent and oceanic boundary and the basin are also referred to. The profiles are divided into seven parts in accordance with the direction of track-lines: (A) Xisha, (B) Continent-ocean boundary, (C) North and (D) Southeastern basin, (E) Zhonsha, (F) Southwestern sub-basin, and (G) Nansha regions

(Fig. 8). The average acoustic basement depth is 6.0 km in North, South, Southwestern basins, 6.0 km in Xisha, 1.5 km of Zhonsha, 2.5 km in Nansha and 3.0 km in the continent-ocean boundary. The gravity anomalies reflect the flexural response of the plate by the topographic loads (e.g., Dorman and Lewis, 1970; Watts and Cochran, 1974; McKenzie and Bowin, 1976; Banks et al., 1977; McNutt and Parker, 1978; Louden, 1981; Forsyth, 1985, Nakasa and Seno, 1994). Details of this technique are given in Walcott (1970-a, -b), and Watts and Talwani (1974).

Seven parts of SCS have following characteristics (Nakasa, 1985).

(A)   Xisha depression;   Eight profiles were obtained in the Xisha depression. Six theoretical calculations give the best fit to a model of 2 km thick elastic plate underlying 8 km thick crust.

(B) Continent-ocean boundary;   The low gravity (as well as MQZ) belt is divided into three segments and each one has different mechanical response.   In the western and the eastern

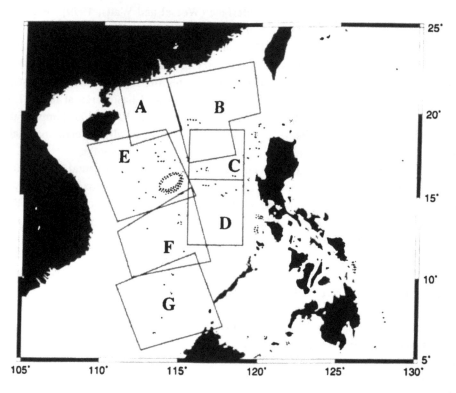

Figure 8.   Gravity dataset are divided into seven parts in accordance with the direction of track-lines: (A) Xisha, (B) continent-oceanic boundary area, (C) northern part of the main basin (refer to Fig. 2), (D) southeastern part of the main basin (refer to Fig. 2), (E) Zhonsha, (F) southwestern sub-basin (refer to Fig. 2), and (G) Nansha regions. These regions are gravimetrically tested regarding their mechanical characteristics.

part, 5 km thick crust is floating isostatically over the mantle. In the central part, however, a compensation with an elastic plate below the crust thinner than 2 km is the most probable case.

(C) North of the SCS basin; Of eight profiles, the best fit to the observed admittance-wave number plot is the model of non-elastic plate underlying crust 5-7 km thick.

(D) South of the SCS basin; Non-elastic model of the plate thickness of 1 km below 5 km thick crust can explain well the observed data of nine profiles.

(E) Zhonsha depression; The best fit to the observed admittance-wave number plot is the model of 2 km thin elastic plate underlying 8 km thick crust.

(F) Southwest SCS sub-basin; Twelve profiles are compared with nine theoretical curves for three cases of non-elastic plate underlying 4-6 km thick crust and six cases of 1-2 km thick elastic plate underlying 4-6 km thick crust.

(G) Nansha Island; From twelve profiles an elastic model of the plate thickness 1 km underlying 5 km thick crust can explain well the observed data.

Most part of SCS seems to be compensated by the Airy-type buoyancy. However, some parts of the basin, i.e., central part of COB, Xisha depression and Nansha and its southern area, seem to be supported by elastic strength of the lithospheric plate though it is far thinner than the geometrical plate thickness. The existence of a thin elastic plate of only a few km is of mechanical definition. We are to carefully consider in each case what the actual compensation mechanism is likely to be.

INTEGRATION OF GEOPHYSICAL DATA

Two-dimensional forward modeling of magnetic and gravity anomalies are performed along the main track line (Fig. 6), parallel to the lines by Hayes (1992). A structural modeling is based on the program by Webring (1985) which compares the magnetic and gravity values with the geological structure. The magnetic data are divided into three parts; the continental part, MQZ, and the oceanic crust. FA shows a sudden change from positive to negative over MQZ. Density data are assigned to seismic formations according to an empirical law for velocity (V: identical to Vp) versus density (Ludwig et al., 1971):

r (density) = $- 0.6997 + 2.2302 \times V - 0.598 \times V^2 + 0.07036 \times V^3 - 0.0028311 \times V^4$.

The crustal thickness gradually thins to the oceanic basin side, less than a half of that of the continental shelf area. COB seems to extend more than 100 km and has high velocity layer at the bottom of the lower crust. The results of the modeling are shown in Fig. 9. Between 80 to 150 km from the shore line, apparently volcaniclastic sediments of fairly strong magnetization are distributed from west to east. It is assumed for simplicity that the oceanic crust underneath MQZ is very weakly magnetized as it was shown previously (section 5). The density and magnetization intensity distribution are changed laterally, although the seismic analysis shows rather homogeneous velocity structure in contrast to the magnetic and gravity information (List 1).

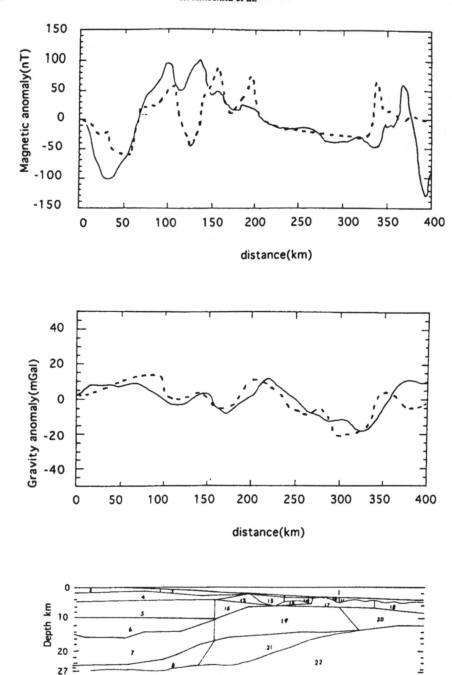

Figure 9. Result of modeling the magnetic and gravity anomaly due to crustal structure (bottom) from the magnetic (top: solid line for observation and dotted line for calculation and gravity (middle: the same as former case applies to the lines) dataset between 0 to 4 km from the shore line. More details are given in the text.

From the continent side to 100 km seaward, a moderately large magnetic anomaly of +/- 100 nT seems to be produced by uppermost sediment and crustal layers. The upper and lower crust ca. 22 km become thinner toward ocean basin. COB extends 100 to 330 km from the shore line above which there exists smooth and thick sediments, non-magnetic solid basement with varying thickness underlain by 10 km thick stretched crust and high velocity layer in the lower crust. The magnetic low might be either due to a non-magnetic layer, or as well due to a pile of several sequences flood basalt each having switching polarity of magnetization. At 330 km away from the shore line, magnetic anomaly Chron 11, the northern end of the lineation group appears. This coincides with the edge of the oceanic crust based on the seismic data.

TECTONIC SETTING OF THE SOUTH CHINA SEA

The Paleogene sedimentary sequences are continental in the northern margin but neritic in the southern margin. This suggests that before the seafloor spreading in 32-17 Ma, the Nansha block remained adjacent to the present northern continental margin. To the south of the Nansha block, there had been paleo-SCS which might have subducted underneath the Sunda arc.

The differences in crustal structure and geophysical characteristics between the northern and the southern continental margins suggest that the seafloor spreading of SCS was asymmetrical. The northern margin experienced both extension in early stage and compression in later stage. The persistent compression in the southern portion of the Nansha block is related to the northward subduction of the Australian plate. If MQZ in the northern margin indicates an existence of an inversely magnetized micro-block, the block might be a part of relict Mesozoic oceanic crust.

The northern shelf is a part of the continental marginal rift system. During Mesozoic time, these rift basins were connected with the South China rift basins. Some of the lower Cretaceous rift basins were formed accompanied by NE faults. MQZ as well as the gravity anomaly, therefore, align mainly along the NE tectonic frame. After Eocene, abrupt spreading along the continental margin occurred.

Activity of subsidence commenced in the south side of the Wanshan fault-uplift (northern SCS), and at the same time, ENE fault movements were activated to create the ENE tectonic background which is reflected partly in the gravity and magnetic anomalies. The whole gravity anomaly was further disturbed by later tectonism, of which a series of NE and NW conjugate faults were sequentially formed. The NW faults were developed to create blocky structures from east to west on the background of ENE and NE tectonic zone. Actually, the blocky gravity anomaly from east to west is observed in the northern shelf. The tectonic evolution in the northern continental shelf is fossilized from north to south and blocky anomaly from east to west.

The BA (refer to p15) is mainly controlled by the deep structure and reflects the depth of the Moho surface. The BA increases gradually with the thinning of the crust from the

Tow-dimensional modelling parameters

Number	region	density(kg/m3)	magnetization(A-m)
1	water	1.030	0.000E+00
2	syn-post sediment	2.150	1.500E-03
3		2.150	1.600E-03
4		1.950	3.550E-03
5		2.420	3.550E-03
6	sediment	2.420	3.550E-03
7	sediment	2.450	1.200E-0.5
8	upper crust	2.820	1.200E-0.5
9	lower crust	3.049	0.000E+00
10	sediment	3.050	0.000E+00
11	sediment	2.200	1.000E-0.7
12	sediment	1.950	0.000E+00
13	upper crust	2.300	0.000E+00
14	sediment	1.900	0.000E+00
15	upper crust	2.050	0.000E+00
16	upper crust	2.000	1.000E-0.7
17	lower crust	2.500	1.300E-0.7
18	lower crust	2.300	1.000E-0.7
19	lower crust	2.150	1.000E-0.6
20	lower crust	2.550	3.000E-03
21	lower crust	2.800	0.000E+00
22	lower crust(oceanic)	2.800	0.000E+00
23	lower crust(oceanic)	3.130	0.000E+00
24	lower crust(oceanic)	3.330	0.000E+00

List 1.   Geophysical parameters of geological formations for modeling of gravity and magnetic anomaly features given in Figure 9. Number of blocks of formations (corresponding to bottom inset of Figure 9), material type, average density of blocks (as described in section 8), and magnetization intensity are given in this list.

northern continental shelf to the Central basin indicating the transition from continental to oceanic crust, and then decreases again toward the southern shelf. The depth of the Moho discontinuity calculated from gravity anomaly reproduces well the seismic structure by refraction experiments. Its thickness (depth measured from seafloor) varies from 31 km to 24 km from continent to the continental margin, 26 km to 13 km across continental slope area, and 13-9 km in the Central basin. The minimal thickness in the Central basin is approximately 5 km.

The late Cretaceous to Eocene ophiolite suits in Sarawak, Sabah, Palawan, Luzon (southern SCS) indicate a formation of late Jurassic to early Cretaceous oceanic crusts in the southern border of the Nansha block (Hutchison, 1968, 1975; Letouzey, 1988). It is possible that this piece of oceanic crust occurred contemporaneously with the piece which carries MQZ in the northern margin. Seismic tomography (Fukao et al., 1994) shows a descending cold slab beneath the China-Indian border through Malay peninsular and Kalimantan to Philippine at the depth of 700-1700 km and another cold slab beneath North and South China at the depth of 2900 km. It shows also that some low velocity (i.e., high temperature) blocks seem to remain near the surface. According to the velocity profiles, there might be a mantle upheaval beneath MQZ. We may assume that there have been an ascending mantle plume beneath SCS. It is possible to consider that the mantle plume results in warming up material beneath the lower crust. The heat flow increases and consequently the velocity and density of the topmost part of the mantle decrease on one hand beneath MQZ. On the other hand, there may have occurred an impingement of the mantle material beneath the lower crust which later solidified to result in a higher velocity formation in the lower crust.

The structure of COB is related to a mode of breakup of the continental crust followed by the seafloor spreading, and then development of the oceanic crust. The comparison between SCS continental margin and the Atlantic margins, both of which are best studied so far, are described briefly. Studies on the Norwegian-Greenland margin have been reported, such as More, Voering plateau (Mutter et al., 1982, 1984, 1988; Eldholm et al., 1989), Lofoten (Shimamura et al., 1994; Goldschmidt-Rokita et al., 1994), southwest Greenland (Chian and Louden, 1994) and southeast Greenland (Goldschmidt-Rokita et al., 1994), Hatton bank (White et al., 1987), Flemish cap (Keen and de Vooga, 1988; Verhoef and Jackson, 1991), Goban spur (Peddy et al., 1989), Scotia margin (Verhoef and Jackson, 1991), Grand banks (Verhoef and Jackson, 1991), and Carolina trough (Holbrook et al., 1994). It was found that there are high velocity layers (7.0-7.5 km/s) underneath the Atlantic margin. COB of SCS has also high velocity layer from 7.0-7.3 km/s under the transition zone. The high velocity layer which has velocity values higher than that of the basement of the oceanic crust but significantly lower than that of upper mantle of surrounding area seems to have been attached to the bottom of the oceanic crust while the continental crust was stretched during rifting phase (White and McKenzie, 1989).

CONTINENTAL RIFT AND OCEAN FLOOR SPREADING

SCS was formed basically by a tension-induced stretching and consecutive rifting between the oceanic and the continental plates. The magnetic anomaly Chrons 11 to 5 (32-17 Ma) of the SCS basin imply that the continuous sea-floor spreading occurred since middle Cenozoic. The north-south opening occurred in the northeastern basin with spreading rate between 22 and 33 mm/yr. It is concluded as follows.

(1) The northern sub-basin consists of a transition zone from continent to oceanic crust

(COB) accompanied by magnetic quiet zone (MQZ).

(2) Part of MQZ is underlain by a seismic high velocity layer in the lower crust by syn-rift impingement.

(3) The age of the northern tip of the oceanic crust is 32 Ma based upon magnetic anomaly age determination (Chron 11).

(4) Continuous sea-floor spreading (Chron 11 to 5: 32 to 17 Ma) is definable over the whole northern basin.

(5) The spreading rate of N-S direction in the northeastern basin fluctuates between 22 and 33 mm/yr.

(6) Faults and fractures divide the spreading axis into several distinct segments.

A weakening or softening of the continental crust before rifting accompanied by ocean floor spreading is a normal sequence of backarc basin formation. Heating, softening then tectonic upheaval and stretching of the continental crust seem to be general consequence for the continental plate to split open. A stretching of the Basin and Range province or an upheaval of Sinai peninsula before rifting are examples of this case (eg. Issler et al., 1989). It is made possible by introducing a heat energy source underneath the continental lithosphere large enough to activate tectonism of the region.

The petrological approach to this problem is accommodated by a lithospheric subduction accompanied by anatexis (high-temperature metamorphism), dragging of hydrated peridotite deeper by subduction, then, dehydration, upwelling of partial melt, formation of wet magma sources (Tatsumi, 1986 and 1989). A thermal plume from dry partial melt from subducted slabs was discussed by a number of authors (e.g., Hasebe et al., 1970; Karig, 1971; McKenzie, 1984). However, a drawback of these theories are that they do not explain adequately the difference between present arc volcanism and past backarc basin opening. Ida (1987) noted that the mantle wedge flow responsible for backarc volcanism is governed by thermal energy of the mantle itself rather than melt of a subducting slab. The mobility of light elements (or volatiles such as water, carbondioxide etc.) reduces effective viscosity by hydrofracturing (e.g., Nakashima, 1993). The water content in the mantle dehydrated from upper part of subducting lithosphere may accelerate the migration of material by reducing melting temperature of minerals (Kushiro et al., 1968). Collision of spreading ridge system might have supplied a good amount of thermal energy and material for rifting and spreading of backarcs. Or a large scale mantle convection as well as advection may play a direct role in the lithospheric splitting followed by its divergence. None of these discussions have completely reached yet a successful conclusion on this problem.

COMMENTS ON THE TECTONIC HISTORY SINCE 40 Ma

Tectonic history of SCS is not fully understood yet. The paleontological and magnetic studies are made in numerous numbers. Ages have been determined for the rock samples recovered from Scarborough seamount chain (Hekinian et al., 1989) and the eastern wall of

the Manila trench where the ridge is being subducted (Pautot et al., 1986; Rangin et al., 1990-a and -b). The formation of SCS proceeded in N-S direction from Oligocene to middle Miocene (32 Ma to 17 Ma) according to the analysis of magnetic lineations of the abyssal basin (Kuramochi, 1993, Briais et al., 1993, Taylor and Hayes, 1980-a and -b, 1983). Taylor and Hayes (1983) suggested that SCS had evolved changing its spreading direction from N-S to NE-SW in the eastern basin. Pautot et al. (1986) suggested that there were two stages of spreading and three rifting episodes. Briais et al. (1989) described that there is possibly an alternative explanation for the E-W and NE-SW trend of magnetic anomalies. Tamaki et al. (1992) suggested that the eastern margin of SCS was initially bounded by a strike-slip fault along the Manila trench and, in general, the basin opened by propagation of the spreading system based on studies of Japan sea (Tamaki, 1995). It is still controversial to settle the opening scheme of SCS due to lack of absolute age of basement
rocks from this area.

The SCS block remained in its present position through Cenozoic era. Major tectonic blocks surrounding SCS remained approximately at their present latitude since early Cretaceous according to the study of the paleomagnetism (Lee, 1988).

Paleocene to Eocene (Rifting and stretching phase)

The Macclesfield bank and the Reed bank remained still in the midst of the proto-SCS. Rifting occurred in NNW-SSE direction around 115°E longitudinal line in late Paleocene to early Eocene. The subsidence of the north of proto-SCS and eruption of suboceanic basalt occurred simultaneously. During the final rifting and early spreading phase, seismic high velocity materials intruded and underplated to the lower continental crust which had been stretched. The Reed bank was formed, and NE-SW trending grabens developed at the Pearl river mouth basin by this spreading activity (Ru, 1988).

Oligocene to Miocene (Spreading phase)

Main part of SCS developed from north to south in 32-26 Ma. The rate of its opening is 2.4 to 2.8 +/- 0.2 cm/yr. Anomaly 7 (25 Ma) trending WNW-ESE is only observable in the northern half of the basin, and indicates that opening of NNE-SSW direction occurred. N-S spreading occurred from 24-21 Ma with a spreading rate ca. 2.9 cm/yr. Anomaly 6c (24 Ma) seems to be oblique but not clearly defined in the eastern part of the basin. The spreading of the NW-SE direction occurred in 21-17 Ma with spreading rate ca. 2.8 cm/yr. The southwestern sub-basin developed during this period.

CONCLUSIONS

A major goal of this study is to describe historical processes of the SCS region in detail. In the oceanic basin part, the newly determined magnetic lineation pattern delineate the trace of the spreading kinematics. As for the pre-spreading stage, COB shows traces of rifting and stretching of continental crust.

The analysis of magnetic, gravimetric and seismic data and other geophysical and geological information from SCS lead to the following conclusions:

(1) N-S direction seafloor spreading started from early Eocene. There were at least four separate evolutional stages. Directions and rates of the spreading are fluctuating and unstable.

(2) The apparent difference in the present tectonism of the eastern and western parts of COB implies that in the east the continental breakup is governed by a strike slip faulting,

(3) The high velocity layer in the lower crust is likely to have been partially underplated to the stretched continental crust.

(4) Spreading of SCS seafloor continued from 32 to 17 Ma.

(5) Magnetic anomaly of the continental margin area seems to be rooted in the uppermost sediment and upper part of lower crust (M-layer hereafter).

(6) The non-magnetic or very weakly and reversely magnetized M-layer is probably responsible for MQZ in COB. One of the causes of demagnetization of the M-layer is due to hydrothermal alteration while high temperature mantle materials being underplated. Another explanation is that a horizontal sequences of basalt each with flip-flop magnetization polarity cancel out the resultant magnetic field on the surface.

DISCUSSION

Studies on the formation of a continental margin play an important role in understanding continental breakup and rifting schemes. Comparison of results of a number of studies from continental margins, such as those in Japan sea, or Atlantic passive margins, with those from SCS reveals tectonic environments of continental margin evolution. Central part of the SCS continent-ocean boundary zone consists mainly of isostatic balance similar to the Atlantic margins (Verhoef and Jackson, 1991). Characteristics of this type has upper crustal faulted and lower crustal stretched. The western part of the SCS margin shows a similar compensation mechanism thus implies a rifted margin. The Xisha depression refers to also the same type of mechanism.

Another type of mechanical balance which shows a sharp gravity contrast across the boundary belt, coincides with a lithospheric margins bounded by transform faults. The eastern part of the SCS margin along the Manila trench is a typical example. This applies to the case of the eastern part of COB of SCS.

A new crustal structure model is produced using a forward modeling of magnetic and gravity with seismic structural data. A quantitative interpretation to estimate the geometry, densities, and magnetization was obtained. The two-dimensional forward calculation results in a new crustal structural model. Because of a variation of density and magnetization to obtain the best fit of modeling, the composition of the crust must be laterally heterogeneous. A new crustal structure model can accommodate thick sediments, non-magnetic basaltic layer (M-layer), stretched crust, and MQZ where the seismic high velocity layer is underplated to the lower crust.

A number of speculations have been discussed. A similar but physically different models were discussed by Hasebe et al. (1970) and by McKenzie (1984). Tatsumi (1986) discussed the backarc volcanism based on dehydration of weathered upper lithospheric materials. Although their results are not complete yet, some basic phenomena for the backarc basin formation can be explained. In addition to stretching and later split of the continental crust, a scattered magnetic lineation pattern is due to scattered multiple magma eruption in a wide area (Tamaki, 1986) which could be explained by introducing a collision of segmented ridges with the continental margin. It is possible that an active oceanic ridge system collided with the Eurasian plate and subducted beneath it and may have caused a backarc opening in the northwestern Pacific. The change (steep increase) of the spreading rate in the Pacific occurred around 35-40 Ma (e.g., Nakanishi et al., 1992); total production of the global crust shows a minimum value at around 25-30 Ma (e.g., Larson and Olson, 1991); change of direction of motion of the Pacific plate relative to hot spot reference frame occurred at 40-42 Ma. Number of backarc basins along western to northern coast of the Pacific region occurred within 60-20 Ma. Collision of segmented ridge system with a continental margin may have been one of the stimuli of discontinuity in the oceanic plate motion.

Acknowledgment

This work has been supported by the Grant-In-Aid for International Scientific Research Program, Ministry of Education, Science and Culture: serial number of funding 04044043 for 1992 and 1993 fiscal years.

Appendix: List of participants to the Japan and China joint study of 1993-1994

(Japan)	Affiliation at the time of survey
Hajimu Kinoshita	Earthquake Research Institute, University of Tokyo
Junzo Kasahara	
Tetsuzo Seno	
Naoshi Hirata	ditto
Sadayuki Koresawa	
Toshinori Sato	
Yukari Nakasa	
Kiyoshi Suyehiro	Ocean Research Institute, University of Tokyo
Hidekazu Tokuyama	
Masanao Shinohara	ditto
Keizou Sayanagi	
Narumi Takahashi	Faculty of Science, Chiba University
Mayumi Sekine	
Seiichi Miura	ditto
Tomoko Tanaka	
Masataka Kinoshita	Department of Marine Sciences, Tokai University
Hideo Kagami	Faculty of Science, Josai University
Kantaro Fujioka	Japan Marine Science and Technology Center

Takashi Yoshida	Okanishi Maito Co. Ltd.
Yukio Watanabe	Service Engineering Co. Ltd
(China)	
Liu Zhaoshu:	South China Sea Institute of Oceanology, Guangzhou
Xia Kanyuan	
Jiang Shaoren	
Zhou Di	
Zhao Yan	ditto
Zhang Yixiang	
Yan Ping	
Wang Ping	
Zeng Guoqing	Captain of research vessel
Zhuang Nong	
Cai Xianheng	ditto
Huang Chengfa	
Yu Zhengguo	Dynamite specialist
Yang Wensheng	Guangzhou Salvage Bureau

REFERENCES

1. G.B. Airy. On the computation of the effect of the attraction of mountain-masses as disturbing the apparent astronomical latitude of station of geodetic surveys, *Phil. Mag. Royal Soc.*, **145**, 101-104 (1855).

2. H. Akima. A new method of interpolation and smooth curve fitting based on local procedures, *J. Acm.*, **17**, 589-602 (1970).

3. H. Akima. A method of bivariate interpolation and smooth surface fitting based on local procedures, *Comm. Acm.*, **17**, 18 - 20 (1974).

4. R.J. Banks, R. L. Parker, and S. P. Huetis. Isostatic compensation on a continental scale: local versus regional mechanisms, *Geophys. J. R. astr. Soc.*, **51**, 431-452 (1977).

5. I.C. Briggs. Machine contouring using minimum curvature, *Geophys.*, **39**, 39-48 (1974).

6. A. Briais, P. Patriat and P. Tapponnier. Reconstruction of the South China Basin and implications for Tertiary tectonics in Southeast Asia, *Earth Planet. Sci. Lett.*, **95**, 307-320 (1989).

7. A. Briais, P. Patrait, P.and P. Tapponnier. Updated interpretation of magnetic anomalies and seafloor spreading stages in the South China Sea: Implications for the Tertiary tectonics of SE Asia. *J. Geophys. Res.* **98 (B4)**: 6299-6328 (1993).

8. S. Chen. The research of the spreading magnetic anomaly and the formation of the South China Sea (1985).

9. B. Chen. Explanation for map of crustal structures. In Atlas of Geology and Geophysics of South China Sea, Map Publishing House of Guangdong Province (1987).

10. X. Chen. The crustal structure in the northern continental margin of the South China Sea as interpreted based on gravity data, in press (in Chinese) (1995).

11. D. Chian and K. E. Louden. The continental-ocean crustal transition across the southwest Greenland margin, *J. Geophys. Res.*, **99**, 9117-9135 (1994).

12. L.M. Dorman and B. T. Lewis. Experimental isostasy, 1, Theory of the determination of the Earth's isostatic response to a concentrated load, *J. Geophys. Res.*, **75**, 3357-3365 (1970).

13. O. Eldholm, J. Thiede and E. Taylor. Voring plateau continental margin: Seismic interpretation, stratigraphy, and vertical movements, *Proc. ODP, Sci. Results*, **104**, 993-1030 (1989).

14. D.W. Forsyth. Subsurface loading and estimates of the flexural rigidity of continental lithosphere, *J. Geophy. Res.*, **90**, 12623-12632 (1985).

15. Y.S. Fukao, M. Murayama, M. Obayashi and H. Inoue. Geologic implication of the whole mantle P-wave tomography. *J. Geol. Sci. Japan*, **100**, 4-23 (1994).

16. A. Goldschmidt-Rokita, J.F.H. Knut, H. B. Hirschleber, T. Iwasaki, T. Kanazawa, H. Shimamura and M. A. Sellevoll. The ocean/continent transition along a profile through the Lofoten Basin, Northern Norway, *Marine Geophys. Res.*, **16**, 201-224 (1994).

17. W.B. Harland, R. L. Armstrong, A. V. Cox, L. Craig, A. G. Smiths and D. G. Smith. *A geologic time scale*, 263 pp (1989).

18. K. Hasebe, N. Fujii and S. Uyeda. Thermal process under island arcs, *Tectonophys.*, **10**, 335-355 (1970).

19. D.E. Hayes. Margins of the southwest sub-basin of the South China Sea - a frontier exploration target ? , Proc. of the Second Workshop on the Geology and Hydrocarbon Potential of the South China Sea and Possibilities of Joint Development, *Energy*, **10**, 373-382 (1988).

20. D.E. Hayes. The tectonic evolution of the greater South China Sea, *Abstract for the 1992 Western Pac. Geophys. Meet.*, Hong Kong (1992).

21. D.E. Hayes, S.S. Nissen, P. Buhl, J. Diebold, Y. Bochu, Z. Weijun and C. Yongqin. Through-going crustal faults along the northern margin of the South China Sea and their role in crustal extension, *J. Geophys. Res.*, **100**, 22435-22446 (1995).

22. R. Hekinian, P. Bonte, G. Pautot, D. Jacques, L. D. Labeyrie, N. Mikkelson and J. L. Reyss. Volcanics from the South China Sea ridge system, *Oceanologica acta*, **12**, 101-115 (1989).

23. K. Hinz and H. U. Schluter. Geology of the Dangerous Grounds, South China Sea, and the continental margin off southwest Palawan : results of Sonne Cruise SO-23 and SO-27, *Energy*, **10**, 297-315 (1985).

24. W.S. Holbrook, E. C. Reiter, G. M. Purdy, D. Sawyer, P. L. Stoffa, J. A. Austin, Jr., J. Oh and J. Makris. Deep Structure of the U. S. Atlantic continental margin, offshore South Carolina, from coincident ocean bottom and multichannel seismic data, *J.*

Geophys. Res., **99**, 9155-9178 (1994).

25. J. Huang and J. Ren. The Chinese geotectonics and its evolution, Science Press, 39-40 (1980).

26. C.S. Hutchison. Tectogene hypothesis applied to the pre-Tertiary of Sabah and the Philippines: *Geological Society of Malaysia Bulletin*, **2**, 65-79 (1968).

27. C.S. Hutchison. Ophiolites in Southeast Asia: *Bull. Geol. Soc. Amer.*, **86**, 61-86 (1975).

28. IAGA Division V Working Group 8 (D. R. Barraclough, Chairman). International Geomagnetic Reference Field, Revision, *J. Geomag. Geoelectr.*, **43**, 1007-1012 (1991).

29. Y. Ida. Structure of the mantle wedge and volcanic activities in the island arcs, High P. Res. Mineral Phys., edit. Maghnani, M.H. and Syono, Y., Terra Publ. Co., Tokyo/AGU, Washington, D.C., 473-480 (1987).

30. N. Isezaki. A new shipboard three-component magnetometer, *Geophysics*, **51**, 1992-1998 (1986).

31. D.Issler, H. McQueen and C. Beaumont. Thermal and isostatic consequence of simple shear extension of the continental lithosphere, *Earth Planet. Sci. Lettr.*, **91**, 341-358 (1989).

32. D.E. Karig. Origin and development of marginal basins in the western Pacific, *J. Geophys. Res.*, **76**, 2542-2561 (1971).

33. C. Ke. The geotectonic characteristics and the evolution of the South China Sea, "Abstract for the Symposium" of The Second *Annual Conference of Ocean Geological Committee of The Geological Society of China*. (1985).

34. C.E. Keen and B. de Vooga. The continent-ocean boundary at the rifted margin off eastern Canada : New results from deep seismic reflection studies, *Tectonophys.*, **7**, 107-124 (1988).

35. J. Korenaga. Comprehensive analysis of marine magnetic vector anomalies, *J. Geophys. Res.*, **100**, 365-378 (1995).

36. T. Kuramochi. The evolutional history of the South China Sea from the magnetic implications, *B. C. thesis, Ryukyu University*, in Japanese (1993).

37. I. Kushiro, Y.Syono and S. Akimoto. Melting of a peridotite nodule at high pressure and high water pressure, *J. Geophys. Res.*, **73**, 6023-6029 (1968).

38. J.L. Labreque. Program MODELF, revised version (1986).

39. R.L. Larson and P. Olson. Mantle plumes control magnetic reversal frequency, *Earth Planet. Sci. Lettr.*, **107**, 437-447 (1991).

40. T.Y. Lee. Tectonic evolution of the South China Sea, 1988 *DELP Tokyo International Symposium*, Tectonic of Eastern Asia and Western Pacific Continental Margin, 66-67 (1988).

41. T.Y. Lee and L. A. Lawver. Tectonic evolution of the South China Sea region, *J. Geol.Soc. China*, **35**, 353-388 (1992).

42. T.Y. Lee, T. Y. and L. A. Lawver. Cenozoic plate reconstruction of the South China Sea region, *Tectonophys.*, **235**, 149-180 (1994).

43. J. Letouzey, L. Sage and C. Muller. Geological and structural map of eastern Asia, The American Assoc. Petrol. Geol. (1988).

44. S. Li. The geomechanics introduction, *Science Press*, 40-41 (1973).

45. Y.Liu. A new type of Diwa region - The introduction of geotectonic evolution in the *South China Sea* , **8:2**, 194-200 (1984).

46. Z. Liu and X. Chen. The distributed feature of the heat flow and the age analysis in the central basin of the South China Sea, *Geoscience*, **2**, 112-121 (1987).

47. Z. Liu, S. Yang, C. Huang and S. Chen. Geological structure of South China Sea and the continental margin spreading, Developments in Geoscience Academia Sinica, 529-539 (1984).

48. K.E. Louden. A comparison of the isostatic response of bathymetric features in the north Pacific Ocean and Philippine Sea, Geophys. *J. R. astr. Soc.*, **64**: 393-424 (1981).

49. W.J. Ludwig, J. E. Nafe and C. L. Drake. Seismic refraction, in *The Sea*, **4**, edited by A. E. Maxwell, pp. 53-84 (1971).

50. O.Matsubayashi and T. Nagao. Compilation of heat flow data in southeast Asia and its marginal seas, In: V. Cermak and L. Rybach (eds.), *Terrestrial Heat Flow and the Lithosphere Structure*, Springer - Verlag, Berlin, Heidelberg, New York, pp. 445-456 (1991).

51. D. McKenzie. The generation and compaction of partially molten rock, *J. Petrol.*, **25**, 713-765 (1984).

52. D. McKenzie and C. Bowin. The relationship between bathymetry and gravity in the atlantic ocean, *J. Geophys. Res.*, **81**, 1903-1915 (1976).

53. M.K. McNutt. Lithospheric flexure and thermal anomalies, *J. Geophys. Res.*, **89 B13**, 11180-11194 (1984).

54. M.K. McNutt and R. L. Parker. Isostasy in Australia and the evolution of the compensation mechanism, *Science*, **199**, 773-775 (1978).

55. J.C. Mutter, M. Talwani and P. L. Stoffa. Origin of seaward-dipping reflectors in oceanic crust off the Norwegian margin by "subaerial sea-floor spreading", *Geology*, **10**, 353-357 (1982).

56. J.C. Mutter, M. Talwani and P. L. Stoffa. Evidence for a thick oceanic crust adjacent to the Norwegian margin, *J. Geophys. Res.*, **89**, 483-502 (1984).

57. J.C. Mutter, W. R. Buck and C. M. Zehnder. Convective partial melting, 1. A model for the formation of thick basaltic sequences during the initiation of spreading, *J.Geophys. Res.*, **93**, 1031-1048 (1988).

58. M. Nakanishi, K. Tamaki and K. Kobayashi. Mesozoic magnetic anomaly lineations and seafloor spreading history of the northwestern Pacific, *J. Geophys. Res.*, **94**, 15437-15462 (1992).

59. Y. Nakasa, A geophysical on the rifting and spreading of the South China Sea, PhD thesis, University of Tokyo (1995).

60. Y. Nakasa and H. Kinosita. A supplement to magnetic anomaly of the Japan Basin, *J. Geomag. Geoeletr.*, **46**, 481-500 (1994).

61. Y. Nakasa and T. Seno. Compensation mechanism of the Yamato Basin, Japan Sea, J. Phys. Earth, 42, 187-195 (1994).

62. Y. Nakashima. Static stability and propagation of a fluid-fluid edge crack in rock: Implication for fluid transport in magmatism and metamorphism, *J. Phys. Earth*, **41**, 189-202 (1993).

63. National Geophysical Data Center, ETOPO-5 bathymetry/topography data, Data Announce. 88-MGG-02, *Natl. Oceanic and Atmos. Admin., U. S. Dep. Commer., Boulder, Colo.* (1988).

64. National Geophysical Data Center, GEODAS CD-ROM worldwide marine geophysical data, Data Announce. 92-MGG-02, *Natl. Oceanic and Atmos. Admin., U. S. Dep. Commer., Boulder, Colo.* (1992).

65. S.S. Nissen, D.E. Hayes, Y. Bochu, Z. Weijin, C. Yongqin and N. Xianpin. Gravity, heat flow, and seismic constraints on the processes of crustal extension, Northern Margin of the South China Sea, *J. Geophys. Res.*, **100**, 22477-22483 (1995).

66. B. Parsons and J. G. Sclater. An analysis of the variation of ocean floor with age, *J. Geophys. Res.*, **82**, 803-827 (1977).

67. G. Pautot, C. Rangin, A. Briais, P. Tapponnier, P. Beuzart, G. Lericolais, X. Mathieu, J. Wu, S. Han, H. Li, Y. Lu and J. Zhao. Spreading direction in the central South China Sea, *Nature*, **321**, 150-154 (1986).

68. C. Peddy, B. Pinet, D. Masson, R. Scrutton, J. C. Sibuet, M. R. Warner,J. -P. Lefort and I. J. Shroeder. Crustal structure of the Goban Spur continental margin, Northeast Atlantic, from deep seismic reflection profiling, *J. Geol. Soc. London*, **146**, 427-437 (1989).

69. C. Rangin, H. Bellon, F. Benard, J. Letouzey, C. Muller and T. Sanudin. Neogene arc-continent collision, in Sabah, Northern Borneo (Malaysia), *Tectonophys.*, **183**, 305-319 (1990-a).

70. C. Rangin and E. Silver. Geological setting of the Celebes and Sulu Seas., In: Rangin, C., E. Silver, M. T. von Breymann, et al., 1990, *Proc. ODP, Init. Repts.*, **124**, College Station, TX, 34-42 (1990-b).

71. K. Ru and J. D. Pigott. Episodic rifting and subsidence in the South China Sea, *Amer. Assoc. Petro. Geol. Bull.*, **70**, No. 9, 1136-1155 (1986).

72. D.T. Sandwell and W. H. F. Smith. Global marine gravity from ERS-1, Geosat, and Seasat reveals new tectonic fabric, *EOS Trans. AGU*, **73**, 133 (1992).

73. SCSIO (South China Sea Institute of Oceanology), Geological Structure and Continental-Margin Spreading in the South China Sea. China Science Press, Beijing, 398p, in Chinese (1986).

74. N. Seama, Y. Nogi and N. Isezaki. A new method for precise determination of the position and strike of magnetic boundaries using vector data of the geomagnetic anomaly field, *Geophys. J. Int.*, **113**, 155-164 (1993).

75. M. Sekine, N. Hirata, H. Kinoshita and T. Seno. Seismic crustal structure of the northern South China Sea by the Japan-China joint study, Abstract for AGU 1994 *Western Pacific Geophysical Meeting* (1994).

76. H.Shimamura, S. Kodaira, H. Shiobara, M. Mochiduki, Y. Nemoto and A. Nakanishi. The initiation of the breakup of the Atlantic ocean using Ocean Bottom Seismographs,in Japanese, *Earth Monthly*, **9**, 76-84 (1994).

77. K. Tamaki. Age estimation of the Japan Sea on the basis of stratigraphy, basement depth, and heatflow data, *J. Geomag. Geoelectr.*, **38**, 427-446 (1986).

78. K. Tamaki, K. Suyehiro, J. Allan, J. C. Ingle and K. A. Pisciotto. Tectonic synthesis and implications of Japan Sea ODP drilling, in *Proc. ODP, Sci. Results*, **127/ 128:2**, 1333-1348 (1992).

79. K. Tamaki. Opening tectonics of the Japan Sea, in Backarc Bains : *Tectonics and Magmatism*, edited by Brian Taylor, pp. 407-420 (1995).

80. Y. Tatsumi. Formation of the volcanic front in subduction zones, *Geophys. Res. Lettr.*, **13**, 717-720 (1986).

81. Y. Tatsumi. Migration of fluids and genesis of basalt magmas in subduction zones, *J. Geophys. Res.*, **94**, 5497-4704 (1989).

82. B. Taylor and D. E. Hayes. The tectonic evolution of the South China Sea Basin, In : Hayes, D.E. (ed.), The tectonic and geologic evolution of Southeast Asian Seas and Islands, *Geophys. Monogr.* Ser.23, Amer. Geophys. Union, Washington,D.C.Part 1, 23-56 (1980-a).

83. B. Taylor and D. E. Hayes. The tectonic evolution of South China Basin. In: D.E.Hayes (ed.), The Tectonic and Geologic Evolution of Southeast Asian Seas and Islands. *Geophys. Monogr.* Ser.23, Amer. Geophys. Union, Washington,D.C.,Part 1, 89-104 (1980-b).

84. B. Taylor and D. E. Hayes. Origin and history of the South China Sea basin. In: D.E.Hayes (ed.), *Geophys. Monogr.* Ser. 27, AGU, Washington, D.C.: 23-56 (1983).

85. J. Verhoef and H. R. Jackson. Admittance signatures of rifted and transform margins : examples from eastern Canada, *Geophys. J. Int.*, **105**, 229-239 (1991).

86. R.I. Walcott. Flexural rigidity, thickness, and viscosity of the lithosphere, *J. Geophys. Res.*, **75**, 3941-3954 (1970-a).

87. R.I.Walcott. Flexure of the lithosphere at Hawaii, *Tectonophys.*, **9**, 435-446 (1970-b).

88. A.B.Watts and J. R. Cochran. Gravity anomalies and flexure of the lithosphere along the Hawaiian-Emperor Seamount Chain, *Geophys. J. R. Astr. Soc.*, **38**, 119-141 (1974).

89. A.B.Watts and M. Talwani. Gravity anomalies seaward of deep-sea trenches and their tectonic implications, *Geophys. J. R. Astr. Soc.*, **36**, 57-90 (1974).

90. A.B.Watts. An analysis of isostasy in the worldℑs oceans 1, Hawaiian-Emperor Seamount chain, *J. Geophys. Res.*, **83**, 5989-6004 (1978).

91. A.B. Watts and M. S. Steckler, Subsidence and tectonics of Atlantic-type continental margins, *Oceanologica Acta*, SP, 143-153 (1981).

92. M. Webring. A FORTRAN program for generalized linear inversion of gravity and magnetic profiles, *U. S. Geol. Surv.*, Open file Rep., 85-122 (1985).

93. J.K.Weissel and A. B. Watts. Tectonic evolution of the Coral Sea Basin, *J. Geophys. Res.*, **84**, 4572-4582 (1979).

94. R.S. White, G. K. Westbrook, A.N. Bowen, S.R. Fowler, G.D. Spence, C. Prescott, P.J. Barton, M. Jopper, J. Morgan and M.H.P. Bott. Hatton Bank (northwest U. K.) continental margin structure, *Geophys. J. R. Astr. Soc.*, **89**, 265-272 (1987).
95. R.S. White and D. McKenzie. Magmatism at rift zone: the generation of volcanic continental margins and flood basalt, *J. Geophys. Res.*, **94**, 7685-7729 (1989).
96. K. Xia and D. Zhou. The geophysical characteristics and evolution of northern and southern margins of the South China Sea by the Japan-China joint study. *Geol. Soc. Malaysia Bull.*, **33**, 223-240 (1993).
97. K. Xia, K., C. Huang, S. Jiang, Y. Zhang, D. Su, S. Xia and Z. Chen. Comparison of the tectonic and geophysics of the major structural belts between the northern and southern continental margins. *Tectonophys.*, **235**, 99-116 (1994).

Proc. 30ᵗʰ Int'l. Geol. Congr., Vol. 13, pp. 33-46
Wang & Berggren (Eds)
©VSP 1997

The Rifting and Collison of The South China Sea Terrain System

JIABIAO LI
Department of Marine Geology and Geophysics, Second Institute of Oceanography, SOA, Hangzhou, 310012, China

Abstract

The Mesozoic northwestward amalgamation and Cenozoic southeastward rifting and collison of the South China Sea terrain system and its influence on the formation and evolution of the South China Sea basin are discussed based on a geophysical data set, dredged rocks and sediment cores obtained during Chinese and joint investigation in the South China Sea, as well as the results of exploration along the margins. Two geological interpretation profiles from north to south of the South China Sea are chosen to illustrate the origin and geodynamic process of the South China Sea.
In the terrain system, Dongsha terrain, Zhongsha-Xisha (Macclesfield Bank-Paracel Is.) terrain, Nansha (Spratley Is.) terrain and North Palawan terrain have a close relation to each other and no relation to South Palawan and Borneo in the late Mesozoic. During Early Cretaceous, these terrains underwent the same tectonic event and constituted an united terrain, although each has its own characteristics.
After the northwestward amalgamation and collison along the South China continental margin in the Late Cretaceous, the united terrain began separating and rifting to the south in the Cenozoic, thus leading to the formation of the South China Sea basin.
The eastern area of the South China Sea basin has the characteristics of seafloor spreading, while the western one has the characteristics of crustal rifting. The rifting separated the Zhongsha-Xisha (Macclesfield Bank-Paracel Is.) terrain, Nansha (Spratley Is.) terrain and North Palawan terrain from north continental margin, and forced them move to the south. Finally in late middle Miocene, the Nansha (Spratley Is.) terrain and North Palawan terrain collided with the Cagayan arc and Sabah-Sarawak terrain along Palawan Trough and solidified. By then, the large scale horizontal movement of terrains stopped.

Keywords: terrain system, amalgamation, rifting, collision

INTRODUCTION

The South China Sea is a great one of the western Pacific marginal-sea basins. It consists of three oceanic-crust subbasins (East subbasin, SW subbasin and NW subbasin) and several thinning continental-crust terrains (Zhongsha-Xisha (Macclesfield Bank-Paracel Is.) terrain, Nansha (Spratley Is.) terrain and others) (Fig.1) [6]. The basin is bordered on the east by an active subduction zone of the Manila Trench, on the west by a NS trending strike-slip fault along the Vietnam margin so called "Vietnam transform fault", on the north by a passive continental margin of mainland China with NE trending grabens and horsts and on the south by a relict subduction zone of Palawan Trough which separates the foreland and forearc area on the north and south [10, 12-13, 15-16].

Because the South China Sea has complicated sedimentary sequences, magmatic activities, strutural styles, crustal composition and tectonic evolution history, several large scale marine geological and geophysical investigations and international cooperative researches have been

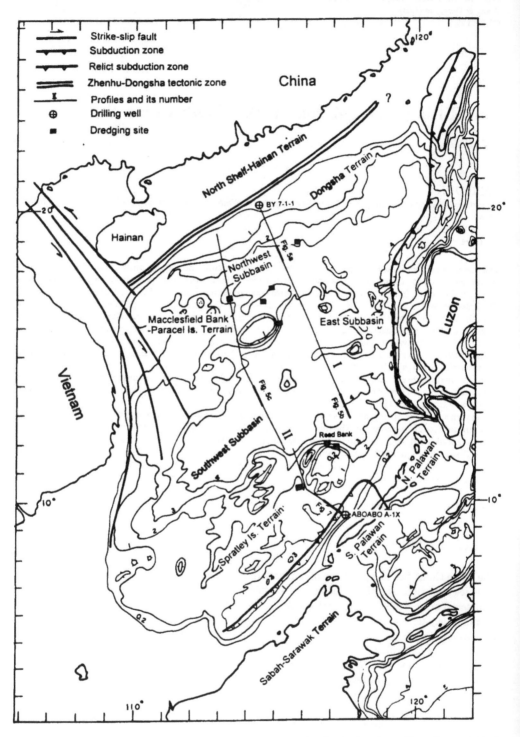

Figure 1   Schematic tectonic map of the South China Sea, showingthe locations of terrains, geological interpretation lines, seismic lines and dredging sites. Isobaths in km. Topography after G.Pautot [10].

carried out by some institutions from China, United States, France and Germany in the last decades [3, 10, 15-16]. The results have improved our knowledge about the South China Sea.

Among all geological problems about the South China Sea, the origin and geodynamic process has long been the focus of many geoscientists, and researchers have put forward a series of models. In these models, the most accepted considers the South China Sea as originating from the break up of the margin of SE Asia [4, 15-16], the second one is the extrusion tectonics of the Indochina block against Asian continent inducing the opening of the South China Sea [14]. Some other authors proposed many other models similar or combined with above one or two and others [3, 10-11].

However, some geological problems, such as the convergent process and features on the north margin before the rifting, exact time when the tectonic regime was changed from convergence to rifting and its reason, difference of tetonic regimes in the east and west part of the South China Sea,and Cenozoic southward convergent process on the south margin, are still unclear and become the matters of debate.

We have combined the results of geological and geophysical cruises of the South China Sea with recent data of exploration along the margins, established two geological interpretation profiles from south to north of the South China Sea and used the method of terrain tectonics to try to illustrate the origin and tectonic evolution history of the South China Sea.

TERRAIN SYSTEM OF THE SOUTH CHINA SEA

Terrain (or terrane), as an analysical method of tectonic evolution, has been extensively used to the studies of continental accretion and different geological block amalgamation since 1980's [1, 9]. According to the features of sedimentary sequences, magmatic and metamorphic activities, structural styles and tectonic evolution, seven terrains in or around the South China Sea were chosen to be studied (Fig.1). They are the North Shelf-Hainan terrain, Dongsha terrain, Zhongsha-Xisha (Macclesfield Bank-Paracel Is.) terrain, Nansha (Spratley Is.) terrain, North Palawan terrain, South Palawan terrain and Sabah-Sarawak terrain. Furthermore, these terrains can be divided into three series which have different tectonic evolution histories: one is North Shelf-Hainan terrain, the second is composed of Dongsha terrain, Zhongsha-Xisha (Macclesfield Bank-Paracel Is.) terrain, Nansha (Spratley Is.) terrain, North Palawan terrain, and the third consists of South Palawan terrain and Sabah-Sarawak terrain (Fig.2).

The North Shelf-Hainan terrain is located in the south continental margin of the Southeast China which has undergone several complicated southeastward accretions, and its southern limit is NEE-strike Shenhu-Dongsha tectonic zone which extends from the southern slope of Hainan to the northern slope of Dongsha Is. (Fig.1). This terrain has Late Cretaceous sedimentary-metamorphic basement and an extensive distribution of Late Cretaceous granitoid rocks. In Hainan, the folded lower Cretaceous strata are uncomformably overlain by sediments with lack of upper Cretaceous and there is an extensive distribution of intermediate-acid intrusive and volcanic rocks which is amounted to about 30% total magmatic rocks in the island. The isotopic ages of these magmatic rocks range from 151.9 Ma to 70 Ma and are mainly concentrated between 110 Ma and 70 Ma [8]. In north shelf of the South China Sea, on other hand, recent exploration has shown that the basement is mainly composed of upper Cretaceous metamorphic rocks from schist to gneiss uncomformably overlain by Paleogene Sediments and extensively intruded by intermediate-acid magmatic rocks of which the ages range from 130 Ma to 70 Ma and are mainly concentrated between 100 Ma and 70 Ma (Fig.3). Both areas all have undergone the same Late Cretaceous magmatic-metamorphic event.

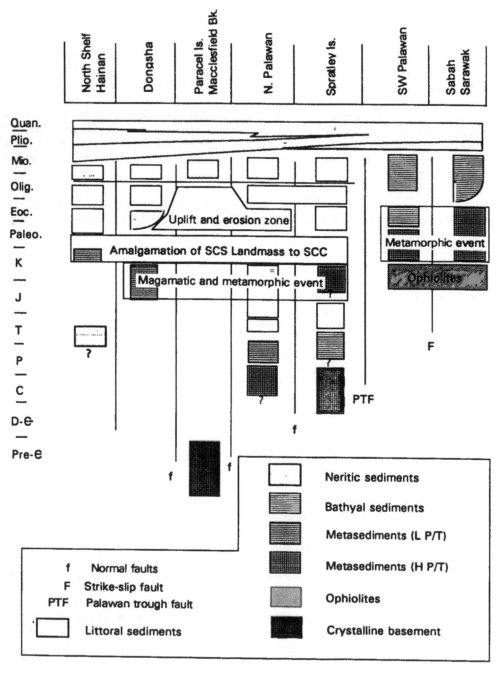

Figure 2 Generalized cross sections of presently contiguous terrains in the South China Sea showing the ages of amalgamation and accretion. L P/T and H P/T present low and high metamorphic pressure-temperature respectively.

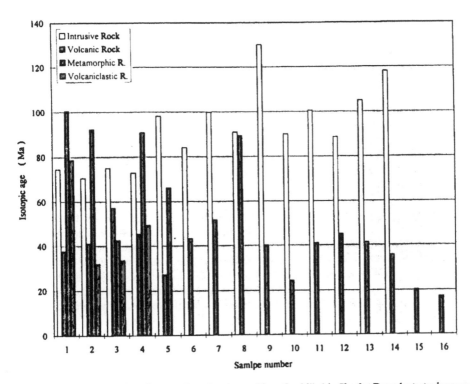

Figure 3 The age distribution of magmatic and metamorphic rocks drilled in Shenhu-Dongsha tectonic zone and its north part of the northern shelf of the South China Sea

The Dongsha terrain has geological feature different from that of North Shelf-Hainan terrain. It's located at the north slope of the South China Sea and has an extensive distribution of intermediate-basic magmatic rocks. Gabbro dredged from KD 17 site is Early Cretaceous and has Sm/Nd age of 134 Ma (Fig.4). On seismic profiles, basement has folding and thrusting reflector structures under the horizontal Cenozoic cover (Fig.5a), and that shows the Dongsha terrain has Early Cretaceous magmatic-folding basement.

Zhongsha-Xisha (Macclesfield Bank-Paracel Is.) terrain has been testified to have a thinning continental crust by geophysical data. At well Xiyong 1 in Xisha (Paracel Is.), highly Pre-Cambrian metamorphic rocks, i.e. granogneiss, was drilled under Neogene sediments. Its Rb-Sr date of whole rock is 627 Ma and rework K-Ar date is 68.9 Ma [17]. During Cruise SO-49 in 1987, several basement rocks were dredged. They are gneiss(KD21), diorite (KD35) and granite (KD27) etc. The granite which was dredged from the eastern flank of Macclesfield Bank has 126.6 Ma and 119 Ma K-Ar dates of mica and plagioclase, respectively (Fig.4). All above show the terrain has an Early Cretaceous magmatic-metamorphic basement which later has undergone Late Cretaceous tectonic event again, and the northern part of this terrain is an intensely erosive area.

The Nansha (Spratley Is.) terrain has the same features as the Zhongsha-Xisha (Macclesfield Bank-Paracel Is.) terrain. During the cruise SO-23 and 27 in 1982-1983, a lot of igneous and metamorphic rocks were dredged from Nansha (Spratley Is.) terrain. Gneiss and phyllite are exposed on the southwest flank of the Reed Band, whose K-Ar ages of muscovite are 122 Ma for the gneiss and 113 Ma for the phyllite, and the K-Ar ages of biotite are 116 Ma and 104 Ma respectively. Metamorphic rocks dredged from the northern end of Reed Bank are 113 Ma K-Ar age of muscovite and 146 Ma K-Ar age of amphibole (Fig.4) [7]. All ages of the dredged rocks belong to Middle Jurassic to Early Cretaceous and mainly Early Cretaceous. Combined with the data of Sampaguita -1 well, this terrain has Early Cretaceous metamorphic-sedimentary basement, and exactly speaking, magmatic-metamorphic basement on the north, sedimentary basement on the south.

The North Palawan terrain has a similar Mesozoic tectonic evolution to the Nansha (Spratley Is.) terrain. Jurassic and Cretaceous rocks have been drilled offshore in the east and west of North Palawan terrain [4, 16]. On the east, the ultrabasic rocks were overlain by poorly sorted quartzoses and stones interbedded with clays and silts of probably Late Cretaceous age, while on the west, upper Jurassic to lower Cretaceous shales and carbonates have a similar sedimentary environment to that of the Reed Bank (Fig.2).

South Palawan terrain and Sabah-Sarawak terrain have similar features in regard to the Mesozoic and Cenozoic tectonic evolution history, but differ from all above other terrains. These two terrains contain similar Early Cretaceous Ophiolites and Late Cretaceous-Paleogene broken formations. In both areas, the Crocker Formation of clastic sedimentary sequences with early middle Eocene (maybe to L. Oligocene) to early middle Miocene has been folded and thrusted northwestwards, and mixed tectonically with poorly dated broken clastic sediments of probably Late Cretaceous-Early Eocene age and dismembered Early Cretaceous Ophiolites partially metamorphosed to amphibolites (Fig.2) [11]

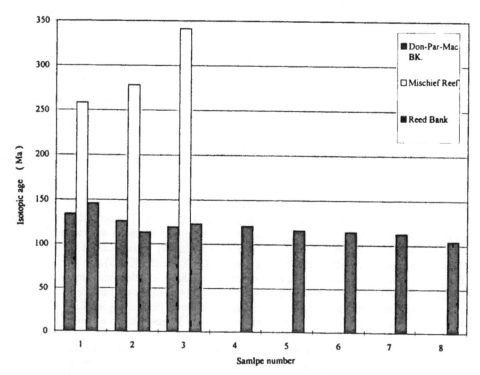

Figure 4 The age distribution of magmatic and metamorphic rocks dredged from Dongsha terrain, Zhongsha-Xisha (Macclesfield Bank-Paracel Is.) terrain and Nansha (Spratley Is.) terrain.   Don-Par-Mac BK.: intrusive rocks from Dongsha terrain and Zhongsha-Xisha (Macclesfield Bank-Paracel Is.) terrain,   Mischief Reef: crystalline source rocks from Mischief Reef,   Reed Bank: metamorphic rocks from Reed Bank.

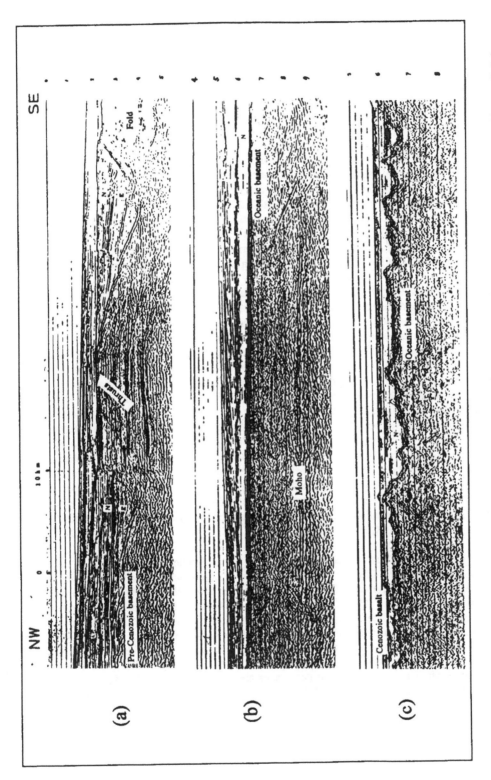

Figure 5 Seismic Profiles in North slope (a), Eastern subbasin (b) and Southwestern subbasin (c) of the South China Sea (for location see Fig.1). N: Neogene, E: Paleogene.

MESOZOIC AMALGAMATION AND ACCRETION OF TERRAINS

Mesozoic tectonic evolution of northern margin of the South China Sea has long been a geological problem of great concern. An Andean-type continent margin of probably Late Mesozoic has been proposed along the northern margin of the South China Sea [4, 15-16], but the exact location and ending time of the Andean-type subduction zone are still a matter of debate [6, 13, 15-16]. However, recent offshore exploration and Hainan geological researches can provide basic constraints on these questions.

Detailed granitoid research in Hainan indicates that Mesozoic granitoid rocks are evidently formed in an active continental margin. These rocks mainly consists of granodiorites and monzogranites. The dates of these rocks, as said above, nearly range from 110 Ma to 70 Ma [8]. Their chemical analyses show that SiO_2 contents range in 59.62-77.24%, AKNC value<1.1 (generally<1.0), KNA value ranges in 0.90-0.50, so these rocks have evolution tendency of classic calcalkaline series. On triangular graphs of AFM and R1-R2, the magmatic evolution tendency is coincident with that in the western margin of northern and southern America [8]. All above indicate that the northern Andean-type margin of the South China Sea has been active until Late Cretaceous.

Between the North Shelf-Hainan terrain and Dongsha terrain, it has been found that there exists an important geological boundary, Shenhu-Dongsha tectonic zone which extends exactly along a large uplift zone of the northern margin of the South China Sea (Fig.1) [6, 18]. Along this tectonic zone, a lot of Mesozoic compressional structures such as reverse fault, thrusts and folds are distributed with basic-ultrabasic rocks. The magmatic and metamorphic evolution on both sides of this boundary is different. On the north, Mesozoic intermediate-acid magmatic series has been considered to be formed on an Andean-type margin with Late Cretaceous ages, and also only Late Cretaceous metamorphic rocks have been drilled, while on the south, rocks of Early Cretaceous intermediate-basic magmatic series have been dredged on Dongsha terrain and metamorphic rocks in the same stage have been drilled at Beigang and Tongliang-1 well in east Dongsha terrain near Taiwan, where the metamorphic strata contain Ammonoidea and Mollusca fauna of Early Cretaceous [6, 18]. This tectonic zone also corresponds an erosive zone of basement where a high angular uncomformity is occurred between the basement and sedimentary cover on the seismic profiles, and an anomalous crust with a thinning upper crust or even lack of all upper crust is directly overlain by sedimentary covers on sonobuoy profiles [2]. So this tectonic zone is probably Late Cretaceous amalgamation and collision zone along the former subduction zone of the Andean-type margin of Southeast China.

In fact, the Southeast China continent has undergone several southeastward complicated accretions of terrains which have different tectonic evolution histories since Caledonian movement. In Yanshanian tectonic stage, a series of magmatic zones have been formed along the southeast margin of the Southeast China and immigrated southeastward with time. An Andean-type magmatic zone of the Late Jurassic-Early Cretaceous is very magnificent along the onshore of Zhejing, Fujian and Guandong province of China, and its isotopic ages range mainly from 150 Ma to 125 Ma. After then at the southeast of this zone, a magmatic zone of Early Cretaceous was formed and primarily consists of the crust-remelting-type granites which come into another magmatic evolution series and their isotopic ages are dropped between 125 Ma and 100 Ma [8]. Recently, the offshore continental shelf drilling has revealed a crust-remelting-type magmatic zone of the Late Cretaceous (110-70 Ma) on the North Shelf-Hainan terrain and at the Shenhu-Dongsha tectonic zone (Fig.3). The temporal and spatial evolution of the tectonics and magmatic activities indicates southeastward terrain accretion history of the South China has been active until late Cretaceous.

In the south of Shenhu-Dongsha tectonic zone, Dongsha terrain, Zhongsha-Xisha (Macclesfield Bank-Paracel Is.) terrain, Nansha (Spratley Is.) terrain and North Palawan terrain all have undergone the same magmatic metamorphic tectonic event of Late Cretaceous, although they are separated from each other. The isotopic dates of the magmatic and metamorphic rocks dredged or drilled in these terrains range from 146 Ma to 104 Ma, and mainly from 125 Ma to 110 Ma (Fig.4). The great difference of magmatic activities characterized by intermediate-basic rocks in these four terrains from that characterized by intermediate-acid rocks in Southeast China including North Shelf-Hainan terrain at the same stage at least indicates above four terrains had no relation to the Southeast China at the that time and probably had formed an united terrain by amalgamation in Early Cretaceous (Fig.2). The existence of the Shenhu-Dongsha tectonic zone just indicates the united terrain was at last amalgamated and accreted to the Southeast China in the Late Cretaceous during the subduction along this zone.

CENOZOIC RIFTING AND COLLISION OF TERRAINS

With the end of the northwestward terrain amalgamations at the end of Mesozoic, the Cenozoic saw the southeastward rifting and collision of terrains. Not until the extrusion of the Indochina block against Asian continent since the Eocene [14], the former Mesozoic continental margin of the Southeast China began large scale rifting southwards with a west edge, the Vietnam transform fault, which is a great right-lateral shear fault and has offset of 160km only from 28Ma to 20Ma [12]. Large amount of grabens and horsts, and three subbasins were formed (Fig.5a). For better awareness of Cenozoic terrain evolution, two geological interpretation profiles from north to south of the South China Sea have been chosen (Fig.6), and the result indicates the South China Sea basin has different rifting regimes on the east and west part.

The NW subbasin is characterized by the oceanic crust. From the existence of Oligocene sediments, it should be the eldest subbasin. But from the existence of the anomalous bodies of low density in lower crust or upper mantle (the gravity Moho is 1.3-1.5km deeper than sonobuoy Moho) and middle to a little high heat flow (79-98 mWm^{-2}, average of 88.5 mWm^{-2}), this subbasin also has strong tectonic activity in the late stage.

The East subbasin is also characterized by the oceanic crust. Its magnetic lineations have symmetrical distribution and trend east to west. The Moho depth calculated from gravity and sonobuoy is shallow at central part (12-13km) and becomes deeper to the north and south (15-16km). Different from the SW subbasin, the basement is very "even" on the seismic profiles (Fig.5b). Above features and high heat flow indicate the East subbasin have newer spreading history (32-17Ma) [15-16].

Although it has an oceanic-crust feature from gravity, magnetic and sonobuoy studies, the SW subbasin crust has a great difference compared with the classic oceanic crust of the world. The heat flow (88-152 mWm^{-2}, average of 108 mWm^{-2}) and water depth are more than that of East subbasin, while the gravity and magnetic anomalies are less than that of East subbasin. The basement is very "rough" on the seismic profiles (Fig.5c) and consists of large amount of SW-trending fault blocks. The gravity Moho is shallow in the central part (10 km) and becomes deeper to the two sides (16 km), but the magnetic lineations is difficult to be identified. These contradictions just indicate the SW subbasin has tectonic regime different from that of the East subbasin.

The studies of gravity and magnetic fields as well as stratigraphy, tectonics and morphology

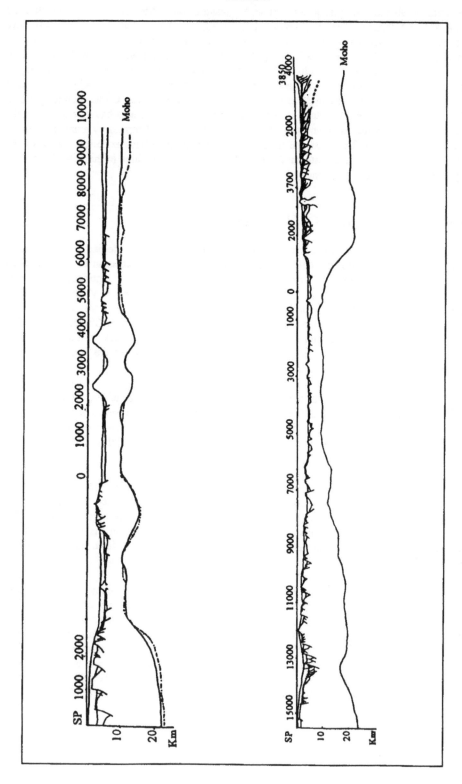

Figure 6 Synthetic geological interpretation of geophysical profiles of profile 1 (upper) and profile II (lower) (for location see Fig.1), showing sedimentary sequence, basement, Moho (heavy line for gravity Moho and dashed line for seismic Moho) and faults

indicate there exists a large scale NWN-trending strike-slip fault in the South China Sea, which extends from Ulugan of the Palawan, through the eastern flank of Reed Bank and Macclesfield Bank, to the north shelf of the South East China continent margin. This fault induces two different tectonic regimes which include microspreading on the east, and crust thinning and rifting on the west.

Cenozoic seismic sequence comparison and seismic facies analysis indicate two structural styles exist in the South China Sea. On the north part of the southwestern end of Reed Bank, the structural style is characterized by grabens and horsts (Fig.5). The uncomformity between lower rifting sequence and upper horizontal sequence is extensively distributed on the South China Sea, and goes at higher and higher levels southwards. The timing of the uncomformity can be determined to be middle middle Miocene by the well ABOABO-1 on the offshore Palawan, thus indicating the southward terrain rifting ended in late middle Miocene.

On the other hand, on the south part of the southwestern end of Reed Bank, the structural style is characterized by foreland folding and thrusting (Fig.7). On the multi-siesmic profiles, the pre-Mid. Miocene strata and basement were intensely overthrusted under the horizontal sequence. These overthrusts mainly developed along the uncomformity or hiatus, which induced the pre-Mid. Miocene strata to be greatly repeated and thickened, and broke them as a series of thrust slices. In the Palawan Trough, this thrusting sequence, together with the thinning crust of Spratley Is., was inserted down under the South Palawan terrain, indicating North Palawan and Nansha(Spratley Is.) terrain rifted southwards and at last collided with the Cagayan volcanic arc, South Palawan and Sabah-Warawak terrain and solidified at the Mid. Miocene (14-15Ma) [11]. The end of Cagayan volcanic arc activity and the emplacement of Palawan and Bukit Mersing Ophiolite zones during the middle Miocene indicate the occurrence of this collision [5, 11].

CONCLUSION

According to the features of sedimentary sequences, magmatic and metamorphic activities, structural styles and tectonic evolution, seven terrains in or around the South China Sea can be divided into three series which have different tectonic evolution histories: one is North Shelf-Hainan terrain, the second is composed of Dongsha terrain, Zhongsha-Xisha (Macclesfield Bank-Paracel Is.) terrain, Nansha (Spratley Is.) terrain, North Palawan terrain, and the third consists of South Palawan terrain and Sabah-Sarawak terrain.

Dongsha terrain, Zhongsha-Xisha (Macclesfield Bank-Paracel Is.) terrain, Nansha (Spratley Is.) terrain and North Palawan terrain have a close relation to each other and no relation to South Palawan and Borneo in Late Mesozoic. During Early Cretaceous, these terrains underwent the same tectonic event and constituted an united terrain, although they are separated from each other now.

The Southeast China continent has undergone several southeastward complicated accretions of terrains which have different tectonic evolution histories since Caledonian movement. In Yanshanian tectonic stage, a series of Andean-type magmatic zones have been formed along the southeast margin of the Southeast China and immigrated southeastward with time.

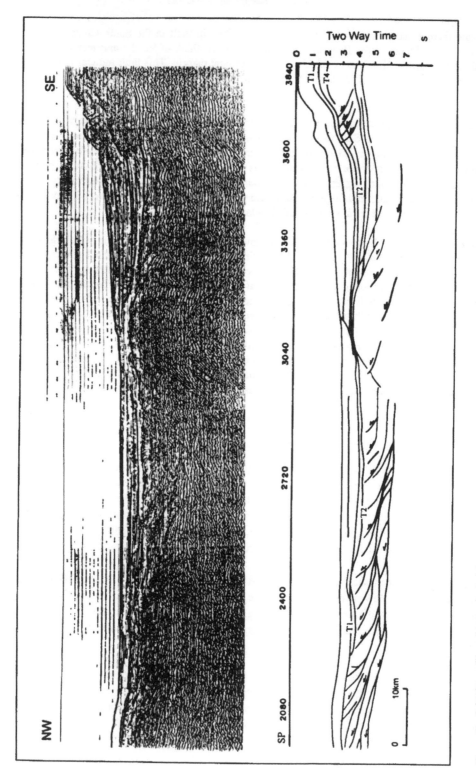

Figure 7 Seismic profile and its geological interpretation in the south of Nansha (Spratley Is.) terrain (for location see Fig.1). T1: the end of U. Miocene, T2: middle M. Miocene, T4: the end of E. Miocene

After the amalgamation and collison of this united terrain along the South China continental margin in the Late Cretaceous, the Dongsha terrain, Zhongsha-Xisha (Macclesfield Bank-Paracel Is.) terrain, Nansha (Spratley Is.) terrain and North Palawan terrain began rifting to the south in Cenozoic, thus leading to the formation of the South China Sea basin.

Cenozoic evolution of the South China Sea basin has two tectonic regimes. On the eastern area of the South China Sea basin, it is characterised by seafloor spreading, while on the western one by crustal rifting from north continental margin to the south.

The rifting separated the Zhongsha-Xisha (Macclesfield Bank-Paracel Is.) terrain, Nansha (Spratley Is.) terrain and North Palawan terrain from north continental margin, and forced them move to the south. Finally in Mid. Miocene, the Nansha (Spratley Is.) terrain and North Palawan terrain collided with Cagayan arc, South Palawan terrain and Sabah-Sarawak terrain along Palawan Trough and solidified. By then, the large scale horizontal movement of terrains stopped.

ACKNOWLEDGEMENTS

This work is a result of "Terrain System Evolution of South China Sea" supported by the National Natural Science Fundation of China and the data are mainly from the Sino-German cooperation in geosciences of South China Sea carried out in 1987-1990. I thank professor X.L. Jin for discussion during the completion of the thesis.

REFERENCES

1. P.J.Coney, D.L.Jones and J.W.Monger. Cordilleran suspect terranes, *Nature* 288, 329-333 (1980).
2. L.He and B.Yao. Sonobuoy measurement on the northern margin of the South China Sea (series 2), *Marine geology and Quaternary geology* 3:4, 57-66 (in Chinese) (1983).
3. K.Hinz and H.U.Schlüter. Geology of the Dangerous Grounds, South China Sea, and the continental margin off Southwest Palawan: results of SONNE cruises SO-23 and SO-27, *Energy* 10:3/4, 297-315 (1985).
4. N.H.Holloway. The stratigraphy and tectonic relationship of Reed Bank, North Palawan and Mindoro to the Asian mainland and its significance in the evolution of the South China Sea. *Bull. Amer. Assoc. Petrol. Geol.* 66, 1357-1383 (1982).
5. C.S.Hutchison. Ophiolite in Southeast Asia, *Geological Society of America Bulletin* 86 797-806 (1975).
6. X.L.Jin. Tectogenesis and origin of Northern South China Sea. In: *Marine geology and geophysics of the South China Sea.* X.L.Jin et al.(Eds). pp.3-9. China Ocean Press (1992).
7. H.R.Kudrass, M.Weidicke, P.Cepek, H.Kreuzer and P.Muller. Mesozoic and Cenozoic rocks dredged from the South China Sea (Reed Bank area) and Sulu Sea and their significance for plate-tectonic reconstructions. *Mar. Pet. Geol.* 3 19-30 (1986).
8. D.Ma et al.. Magmatic rocks. In: *Hainan Geology.* X.Wang et al.(Eds). pp.3-163 Geological Press (in Chinese) (1991).
9. A.Nur and Z.Ben-Avraham. Break-up and accretion tectonics. In: *Accretion Tectonics in the Circum-Pacific Regions.* M.Hashimoto and S.Uyeda (Eds). pp.3-18. Terra Scientific Publishing Company, Tokyo (1983).
10. G.Pautot. Morphostructural analysis of the central ridge in South China Sea. In: *Marine*

geology and geophysics of the South China Sea. X.L.Jin et al.(Eds). pp.10-20. China Ocean Press (1992).

11. C.Rangin. South-East Asian marginal basins (South China, Sulu and Celebes Seas): New data and interpretations. In: Marine geology and geophysics of the South China Sea. X.L.Jin et al.(Eds). pp.38-51. China Ocean Press (1992).

12. D.Roques, C.Rangin, P.Huchon and G.Marquis. Un décrochement dextre NS au long de la marge vietnamienne: implications pour l'ouverture ed la mer de Chine méridionale. Journées spécialisées de la Société Géologique de France, Géosciences Marines, 18-19 Decembre 1995, Brest, France, Abstract, p. 62-63 (1995).

13. K.Ru and J.D.Pigott. Episodic rifting and subsidence in the South China Sea. American Association of petroleum Geologists Bulletin. 70, 1136-1155 (1986).

14. P.Taponnier, G.Peltzer, A.Y.Le Dain, R.Armijo and P.Cobbold. Propagating extrusion tectonics in Asia: New insights from simple experiments with plasticine. Geology 10, 611-616 (1982).

15. B.Taylor and D.E.Hayes. The tectonic evolution of the South China Basin. The Tectonic and Geologic Evolution of Southeast Asian Seas and Islands. Geophys. Monog. D. E. Hayes (Eds). AGU, 27, 89-104 (1980).

16. B.Taylor and D.E.Hayes. Origin and history of the South China Sea Basin. The Tectonic and Geologic Evolution of Southeast Asian Seas and Islands. Geophys. Monog, D. E. Hayes (Eds). AGU, 23-56 (1983).

17. Y.Yan. On the longitude trending tectonic zone of the South China Sea, Geodynamics 4 p.417 (in Chinese) (1981).

18. Yang Shukan, Liu Zhaoshu, Chen Senqiang, Zhou Yan, Zhang Yixiang and Chen Hanzhong. Basement structure and evolution of Pearl River Mouth Basin. Proceedings of Symposium on Petroleum Geology of Northern Shelf in South China Sea Part1, 250-272 (1987).

Proc. 30th Int'l. Geol. Congr., Vol. 13, pp. 47-56
Wang & Berggren (Eds)
©VSP 1997

The Evolution of Marginal Seas off China

XIANGLONG JIN

Second Institute of Oceanography, SOA, Hangzhou, Zhejiang, 310012 China

Abstract

The Japan Sea, East China Sea and South China Sea as the marginal seas between Western Pacific and Eurasian continent are adjacent to China. Each marginal sea is composed from several marginal basins. The history of marginal sea is the evolution history of marginal basins which constituted marginal sea.

The marginal basins are caused by the secondary spreading, and the secondary spreading is generated by the secondary convection due to sliding friction and heating in the process of the descending of oceanic plate as collision. The formation of marginal basin is always accompanied with the transformation of crust. There are three periods of secondary spreading in the South China Sea, the oldest spreading (Cretaceous) formed the Northern Shelf Basin of the South China Sea, the secondary spreading with trend about NE85° produced the Northwestern Subbasin (about Oligocene), the secondary spreading trending near NE70° and NE48° formed the Eastern Subbasin and the Southwestern Subbasin (Oligocene to Miocene); twice secondary spreading in the East China Sea, one caused the East China Sea Basin (Cretaceous to Eocene/Miocene), another — the Okinawa Trough (Pliocene); two times of secondary spreading in the Japan Sea, one produced Japan Basin (late Cretaceous), one — Yamato Basin (Miocene).

The Okinawa Trough is an embryonic marginal basin; the Japan Basin, the Yamato Basin, the Northwestern Subbasin of South China Sea (SCS), the Eastern Subbasin and the Southwestern Subbasin of SCS are mature marginal basins; the East China Sea Basin and Northern Shelf Basin of SCS are senescent marginal basins.

Keywords: South China Sea, East China Sea, Japan Sea, marginal basin, secondary spreading

INTRODUCTION

The submarine geosciences or marine geology and geophysics are especially concerned with the continental margin which consists of the continental shelf, the continental slope and the continental rise. The marginal seas cover most of continental shelf and part of continental slope, that is, the main part of the continental margin. In the past, people regarded the seashore as the margin of continent, and modern geoscientists are beginning to take a more comprehensive view of continental margin that encompasses a wide transitional zone which separates the oceanic from the continental realms. Thus, the continental margins, especially the marginal seas, have now become the joint concern of geoscientists working on the land and their colleagues who work in the oceans.

There are three marginal seas off China between Western Pacific (Philippine Sea) and Eurasian Continent, that is, the South China Sea , the East China Sea and the

Japan Sea, which are adjacent to China. Each marginal sea is composed by several marginal basins. The evolution process of marginal basins which constituted the marginal sea is the geological history of marginal sea.

Study on marine geology of marginal seas and their marginal basins is helpful in understanding the formation and the evolution of marginal seas, marginal basins, even continental margin.

SOUTH CHINA SEA

The South China Sea (SCS) surrounded by China Mainland, Indochina, Borneo and Philippines, is located between three plates, the Eurasian, the Pacific and the Australian plates. The South China Sea consists of several marginal basins and a system of terrains which is composed of the Xisha-Zhongsha (Paracel-Macclesfield) Terrain, the Nansha (Reed bank and Spratly) Terrain with continental crust [9]. (Fig.1)

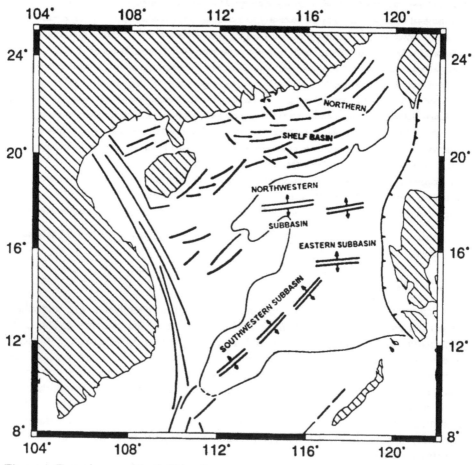

Figure 1. Tectonic map of South China Sea.

The marginal basins in the South China Sea consist of the Northern Shelf Basin and the South China Sea Basin. The Northern Shelf Basin of continental crust is the oldest marginal basin in the South China Sea that connected with the East China Sea Basin. The South China Sea Basin (SCS Basin) is divided into three subbasins: the Northwestern Subbasin north of Xisha (Paracel Islands), the Eastern Subbasin and the Southwestern Subbasin (Fig.1). The SCS Basin is a deep sea basin characterized by deep water (generally more 3000m deep) and oceanic crust beneath it.

The depth of Mohorovicic discontinuity (MOHO) beneath the SCS was calculated by inversion from gravity anomalies [21], and taken from seismic measurement [1, 8]. The distribution of depth to MOHO (Fig. 2) indicates that the crust of the Shelf in the SCS is continental in nature and the depth to MOHO is about 24-18 km, the crust of Continental Slope is transitional with a depth down to MOHO of 24-14 km, and the crust of the SCS Basin is oceanic with the depth to MOHO of 14-10 km and the minimum of 9 km as evidenced by seismic reflection (Fig. 3). The crust beneath terrains in the SCS thickens, the depth to MOHO is 24-22 km and more in Xisha-Zhongsha Terrain, 24 km and more in Nansha Terrain [9].

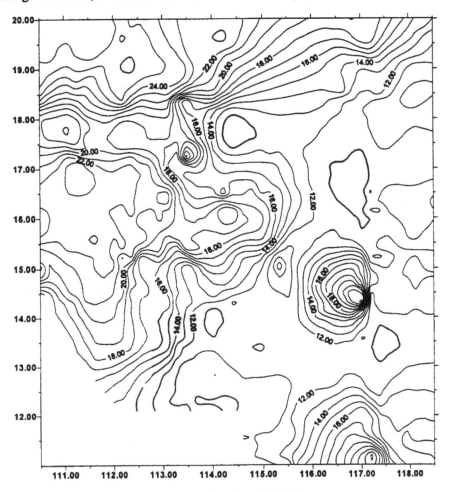

Figure 2. Map of MOHO depth in South China Sea. Unit in km.

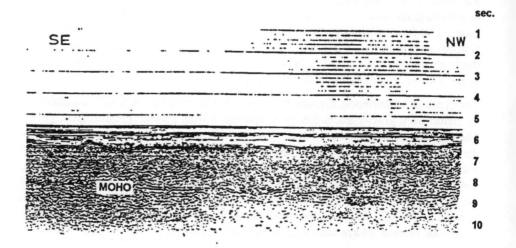

Figure 3. Seismic profile, showing MOHO discontinuity in South China Sea. Unit in sec.

The magnetic anomalies in South China Sea Basin characterized by linear pattern show that the Southwestern Subbasin with magnetic lineation of NE85° is a microspreading but premature failure basin of about Oligocene, the Eastern Subbasin with lineation of NE70°-90° has two spreading episodes of 32-26 Ma (anomalies 11-3) and 26-17 Ma (anomalies 7-5d) from middle Oligocene to early Miocene, and the Southwestern Subbasin with lineation of NE48° is a spreading basin of 20-16 Ma (anomalies 6-5c) in early Miocene [12]. The Northern Shelf Basin with thick sediments could be an oldest spreading basin (Cretaceous) in the South China Sea.

EAST CHINA SEA

The East China Sea (ECS) is a marginal sea between the Asian Mainland and the Western Pacific (Philippine Sea). Most of the East China Sea (the western part) is a part of the vast Asian Continental Shelf. The broad shelf of the ECS is bordered in the east by the Okinawa Trough, a back-arc and deep-water basin. The Ryukyu Islands are distributed along the edges of the ECS between the Okinawa Trough and the Philippine Sea in form of an arc islands. There are two marginal basins in the ECS, the East China Sea Basin (ECS Basin) and the Okinawa Trough.

The calculation from Bouguer anomalies [5] shows that the crust beneath the ECS Basin is continental with thickness of 30-33 km, the crust of northern Okinawa Trough is continental and 20-28 km thick, and the crust beneath the southern Okinawa Trough is transitional and 15-21 km thick which is identical to the results obtained by Lee *et al*. [11]. The thickness of crust is 27-30 km in the Ryukyu Islands, and the crust east of it is oceanic with thickness of 7-9 km in the Ryukyu Trench.

The ECS Basin is a Cenozoic sedimentary basin where thickness of the Cenozoic sedimentary series generally exceeds 4 km with maximum of about 9 km [3]. The lower series of Cenozoic in ECS Basin is the tectonically deformed Eocene and Oligocene deposits with seismic velocity of 4.72-5.67 km/sec, the middle series is the deformed Miocene with velocity of 3.67-4.59 km/sec, the upper series is the slightly folded Pliocene with velocity of 2.51-3.13 km/sec [6]. The East China Sea Basin is divided into three depressions, the northern (Fujiang) depression with sedimentary series of 3-4 km, the middle (Zhedong) depression with thickness of sediments about 7 km and the southern (Taibei) depression with thickness of sedimentary series 7-9 km [3] (Fig. 5). During the Cretaceous to Eocene, the ECS Basin was a rift zone, and the sedimentary basins have formed on this basis, with the eastern part of the ECS Basin formed in late Cretaceous to Eocene, and the western part of the Basin formed from late Cretaceous to Miocene.

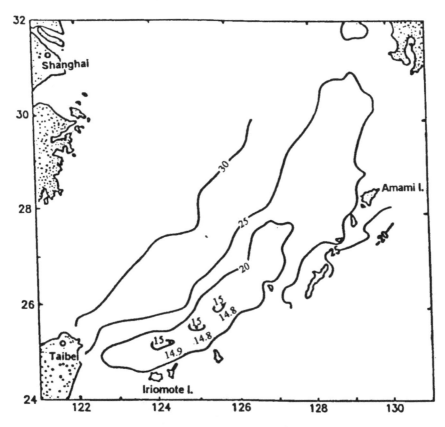

Figure 4. Map of MOHO depth in Okinawa Trough. Unit in km.

The Okinawa Trough as a new marginal basin developed on the basis of continental lithosphere has certain features of transitional structures — the thinned crust, the active volcanism and tectonism, as well as the high heat flow. Three main sedimentary series in Trough have been distinguished by seismic prospecting: the horizontal or sub-horizontal layer A of Holocene-Pleistocene thinner (500-350m) in north and thicker (1700m) in south, the slightly folded layer B of Pliocene thicker (>2000m) in the north and thinner (<1400m) in the south, and the deformed layer C of Miocene below which there may be an another series of Mesozoic-Paleogene deposits being difficult to distinguish from layer C. The Okinawa Trough began its rifting phase in late Miocene, producing a series of rifting basins along the axis of Trough from Miocene to Early Pliocene. The Trough has entered the spreading phase in late Pliocene, when the southern section of Trough began to spread. [7]

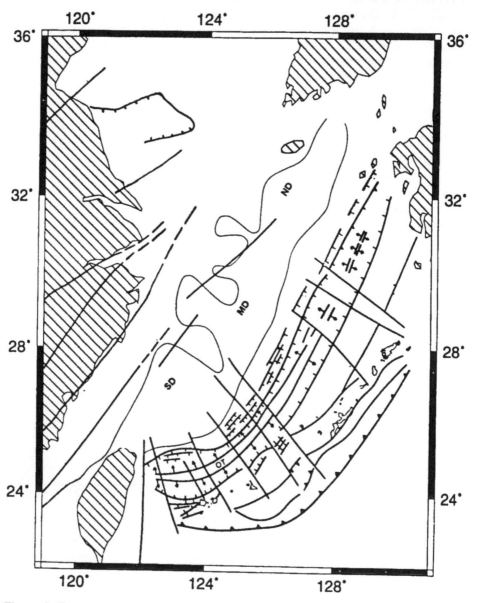

Figure 5. Tectonic map of East China Sea. OT-Okinawa Trough, ND-Northern Depression, MD-Middle Depression, SD-Southern Depression.

JAPAN SEA

The Japan Sea between the Japanese Islands and the Asian Mainland is a marginal sea with two marginal basins separated by the Oki Bank-Yamato Ridge: the Japan Basin and the Yamato Basin. The Japan Basin including the Tsushima Basin is in the north and the west of Japan Sea, and the Tsushima Basin near the Honshu of Japan is in the south of Japan Sea. The crust beneath the Japan Basin and the Tsushima Basin is oceanic [13] or suboceanic [14, 17], the thickness of the crust is 8-9 km in the Japan Basin and is 11-13 km in the Yamato Basin [13], but another scientists such as Minato and Hunahshi [14], Shilo *et al.* [17] believe the crust thickness in the Japan Sea is about 15-18 km.

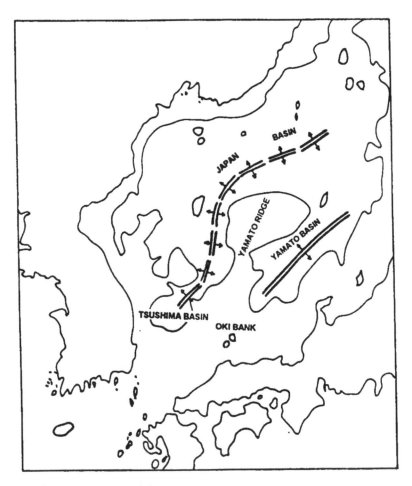

Figure 6. Tectonic map of Japan Sea.

The magnetic anomalies patterns have been distinguished in the Japan Basin with their age assigned to 13-24 Ma. The older anomaly (22-24 Ma) with a half spreading rate of 8.0 cm/yr is in the eastern part of Japan Basin, the magnetic stripes of 22-24 cm/yr and a slower half spreading 7.5 cm/yr in the central part of the basin, and the youngest anomalies of 13-15 Ma with faster half spreading rate 10.0 cm/yr fit to the western part of basin [15]. That means the Japan Basin formed in Miocene. However, Uyeda and Miyashiro [19] suggest that the Japan Basin originated between 80 Ma and 90 Ma, in late Cretaceous. Uyeda [20] also suggests the Yamato Basin formed at about 25 Ma (Miocene), but it seems to be younger than the Japan Basin. Nakasa and Seno [16] hold that the Yamato Basin has a compensation mechanism of elastic plate with flexure.

The Yamato Ridge could be a remnant arc formed while the Japan Basin opened, it has been composed of sedimentary series, granitoid, basaltoid, tuff and crystalline metamorphic series [17]. The crust is continental, and its thickness is about 15-20 km [2].

EVOLUTION OF MARGINAL BASIN

The marginal basins have been classified by Karig [10] into three types: the active basin, the inactive with high heat flow basin, and the inactive with normal heat flow basin. Toksoz and Bird [18] have proposed four stages of evolution of marginal basins: undeveloped, active spreading, mature, and inactive. In the above classifications, the active basins are the same as active spreading basin, the inactive with high heat flow basin is the mature basin, and Toksoz and Bird [18] divided the inactive with normal heat flow basin into two types of basins, that is, undeveloped and inactive. The present author suggests three types of marginal basins or three stages of evolution: the embryonic (initial) marginal basin, the mature (robust) marginal basin, and the senescent (extinct) marginal basin.

The marginal Basins are caused by the secondary spreading, which is generated by the secondary convection due to sliding friction and heating in the process of the subduction of oceanic slab as collision [4]. The formation and evolution of marginal basins is always accompanied with the transformation of crust in its properties. The crust beneath the continental margin began its change from continental to oceanic with the opening of marginal basin due to heating, rifting and thinning of the crust, and the crust turned into oceanic when the marginal basin formed. After the marginal basin has been formed, the crust beneath the marginal basin would become back gradually into continental from oceanic owing to the giant accumulation of sediments and either cease or pause of plate subduction, and the continental shelf will be formed, then the continent grows and expands in consequence of its reverse transformation of crust.

The marginal sea would be formed and expanded by multiple secondary spreading in sequence of time. The history of marginal sea is the evolution of marginal basins in it. Three periods of secondary spreading in the South China Sea have been recognized: the oldest secondary spreading (Cretaceous) caused the Northern Shelf Basin of the South China Sea, the second one produced the Northwestern Subbasin north of Xisha (about Oligocene), and the last secondary spreading formed the Eastern and the Southwestern Subbasins (Oligocene to Miocene). Twice secondary spreading formed the East China Sea, one caused the East China Sea Basin (late Cretaceous to Eocene in the west and to Miocene in the east), another - the Okinawa

Trough (Pliocene). There are two times of secondary spreading in the Japan Sea, producing the Japan Basin (late Cretaceous to Miocene), and Yamato Basin (Miocene), respectively.

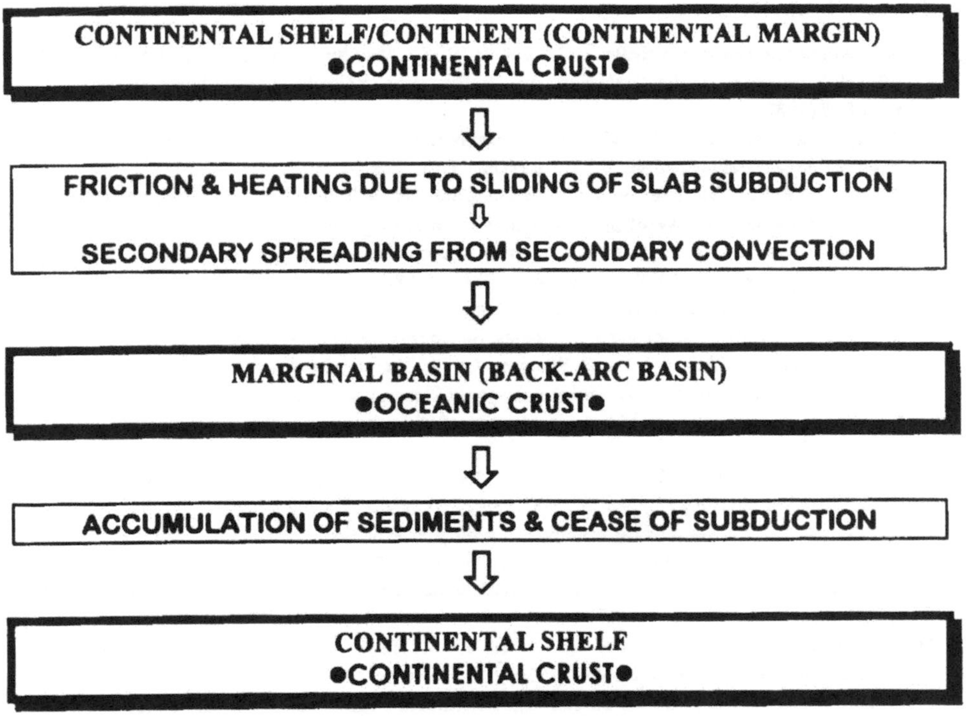

CONTINENTAL SHELF/CONTINENT (CONTINENTAL MARGIN)
●CONTINENTAL CRUST●

FRICTION & HEATING DUE TO SLIDING OF SLAB SUBDUCTION
⇩
SECONDARY SPREADING FROM SECONDARY CONVECTION

MARGINAL BASIN (BACK-ARC BASIN)
●OCEANIC CRUST●

ACCUMULATION OF SEDIMENTS & CEASE OF SUBDUCTION

CONTINENTAL SHELF
●CONTINENTAL CRUST●

Figure 7. Transformation of crust with evolution of marginal basins

Table 1. Classification of marginal basins in stages of evolution

EMBRYONIC	MATURE	SENESCENT
Okinawa Trough	Northwestern Subbasin of SCS	East China Sea Basin
	Eastern Subbasin of SCS	North Shelf Basin of SCS
	Southwestern Subbasin of SCS	
	Japan Basin	
	Yamato Basin	

The marginal basins in the marginal sea off China are classified in stages of its evolution as follows: the Okinawa Trough is an embryonic marginal basin; the Japan Basin and the Yamato Basin, as well as, the Northwestern Subbasin, the Eastern Subbasin and the Southwestern Subbasin of the SCS are mature marginal basins; and the East China Basin with the Northern Shelf Basin of SCS is a senescent marginal basin.

Acknowledgments

My sincere appreciation is expressed to Mrs. Yingxia Fan and Mr. Ziying Wu for their help in drawing by computer.

REFERENCES

1. Bundesanstalt für Geowissenschaften und Rohstoffe (BGR). Abschlussbericht uber die SONNE-Fahrt SO-49, Teil 2, Geophysikalische, Geologische und Geochemische Untersuchungen im Sudchinesischen Meer. pp. 17-18 (1987).

2. Y. Fujita and Y. Ganzawa . The origin and development of the Sea of Japan. In: *Geology of Japan Sea.* Hoshino and Shibasaki (Eds). pp. 37-58. Tokai Univ. Press, Tokyo (1982).

3. X.L. Jin and P. Yu . The tectonics of the East China Sea and the Yellow Sea. In: *Geology of East China Sea and Yellow Sea* (in Chinese). pp. 1-22. Science Press, Beijing (1982).

4. X.L. Jin. The function of marginal basin in growth of continent. In: *Geology of Japan Sea.* Hoshino and Shibasaki (Eds). pp. 143-150. Tokai Univ. Press, Tokyo (1982).

5. X.L. Jin, P. Yu, M. Lin, Ch. Li and H. Wang. Preliminary study on the characteristics of crustal structure in the Okinawa Trough. *Oceanologia et Limnologia Sinica* (in Chinese). 14:2, 105-116 (1983).

6. X.L. Jin, B. Tang, J. Zhuang, J. Yu, G. Wang and Q. Du. The velocity structure of upper crustal layer in East China Sea. *Oceanologia et Limnologia Sinica* (in Chinese). supplement, 65-72 (1986).

7. X.L. Jin and P. Yu. Structure and tectonic evolution of Okinawa Trough. *Scientia Sinica* (ser. B). 31:5, 614-623 (1988).

8. X.L. Jin, W. Lu, Ch. Ke, Q. Li, J. Qian, J. Liu, J. Jiang and Ch. Zeng. Report on study of geosciences in South China Sea. Jin (Ed.). *Donghai Marine Science* (in Chinese) 7:4, 1-92(1989).

9. X.L. Jin and Ch. Ke. Geophysical features and geological evolution of the South China Sea. In: *Asian Marine Geology.* pp. 147-155. China Ocean Press, Beijing (1990).

10. D.E. Karig. Origin and development of marginal basins in the Western Pacific. *JGR.* 76, 2542-2561 (1971).

11. C.S. Lee, G.G.Jr. Shore, L.D. Bibee, R.S. Lu and T.W.C. Hilde. Okinawa Trough: origin of a back-arc basin. *Marine Geology* 35:1-3, 219-241 (1980).

12. W. Lu, Ch. Ke, Sh. Wu, J. Liu and Ch. Lin. Characteristics of magnetic lineations and tectonic evolution of the South China Sea Basin. *Acta Oceanologica Sinica* 6:4, 577-588(1987).

13. W.J. Ludwig, S. Murauchi and R.E. Houtz. Sediments and structure of the Japan Sea. *GSA Bull.* 86, 651-664 (1975).

14. M. Minato and M. Hunahashi. Crustal structure of the deep basin in the Japan Sea. In: *Geology of Japan Sea.* Hoshino and Shibasak (Eds). 21-32. Tokai Univ. Press, Tokyo (1982).

15. Y. Nakasa and H. Kinoshita. A supplement to magnetic anomaly of the Japan Basin. *Jour. of Geomagnetism and Geoelectricity* 46:6, 481-500 (1994).

16. Y. Nakasa and T. Seno. Compensation mechanism of the Yamato Basin, Japan Sea. *Jour. of Physics of the Earth* 42:2, 187-195 (1994).

17. N.A. Shilo, Yu.A. Kosygin, I.I. Bersenev, A.V. Zhuravlyov, R.G. Kulinich, K.F. Sergeyev, I.K. Tuezov and I.I. Khvedchuk. The structure and history of geological evolution of earth's crust in the Japan and Okhotsk Sea regions. In: *Geology of Japan Sea.* Hoshino and Shibasaki (Eds). pp. 59-80. Tokai Univ. Press, Tokyo (1982).

18. M.N. Toksoz and P. Bird. Formation and evolution of marginal basins and continental plateaus. In: *Island Arcs, Deep Sea Trench and Back-arc Basins.* Talwani and Pitman III (Eds). M. Ewing ser. 1 AGU. pp. 379-393 (1977).

19. S. Uyeda and A. Miyashiro. Plate tectonics and Japanese Islands. *Bull. GSA* 85, 1159-1170 (1974).

20. S. Uyeda. Some basic problems in the trench-arc-back arc system. In: *Island Arcs, Deep Sea Trench and Back-arc Basins.* Talwani and Pitman III (Eds). M. Ewing ser. 1 AGU. pp. 1-14 (1977.).

21. D. Xu and J. Jiang. Moho pattern and deep structure in the northern and central parts of South China Sea. *Donghai Marine Science* (in Chinese) 7:1, 48-56 (1989).

Proc. 30th Int'l. Geol. Congr., Vol. 13, pp. 57-64
Wang & Berggren (Eds)
©VSP 1997

Onset of North Atlantic Deep Water 11.5 million years ago triggered by climate cooling

WUCHANG WEI AND ALYSSA PELEO-ALAMPAY
Scripps Institution of Oceanography, University of California, San Diego, California 92093-0215, USA

Abstract

The vital role of North Atlantic Deep Water (NADW) in driving high-amplitude climate change in the last 200 k.y. has been well recognized, but little is known about the early history of NADW. Here we present sedimentological and paleontological data from the North Atlantic and the Norwegian-Greenland Sea and show the onset of NADW at ~11.5 Ma. This postdates the expansion of the Antarctic ice sheet by 2-3 my and thus cannot be the cause of the latter event as previously thought. Current data suggest that the onset of NADW may have been triggered by climate cooling rather than subsidence of the Greenland-Scotland Ridge as generally assumed. Significant climate warming in the near future may reverse the process and shut down NADW, causing a fundamental change in climate.

Keywords: North Atlantic Deep Water, Norwegian-Greenland Sea, paleoceanography, paleoclimate, Miocene

INTRODUCTION

The relatively warm, saline North Atlantic surface water flows into the Norwegian-Greenland Sea, releasing a large amount of heat on cooling. The resulting denser water sinks and overflows three sills on the shallow ridge connecting Greenland and Scotland (the Greenland-Scotland Ridge). The overflows entrain resident North Atlantic water in the course of their descent, and join together to form North Atlantic Deep Water (NADW), which moves southward and ultimately spreads into the South Atlantic, Indian Ocean, and Pacific Ocean. NADW plays an important role in the ventilation of the world's deep oceans, distribution of nutrient contents of deep water, and redistribution of planetary heat budget. Large fluctuations of atmospheric CO_2 and climate in the late Quaternary have been linked to variations in NADW [8, 27, 9, 7, 30, 28]. It is thus important to understand the history of NADW.

A basic question as to when and what triggered the onset of NADW has been unclear. Berger (1972; 1976) assumed the onset of NADW in the late Miocene based on the drastic fall of the CCD in the Central Atlantic [3, 4]. More recent age estimates for the onset of NADW varied from ~34 Ma to 12.5 Ma (all ages in this article are given in the time scale of Cande and Kent, 1995) [10] based largely on carbon isotope data from middle to low latitudes [32, 5, 21, 38]. The trigger for the onset of NADW is generally assumed to be the subsidence of the Greenland-Scotland Ridge [32, 5, 17, 21, 38, 37].

In order to help determine the timing and investigate the trigger for the initiation of NADW, we have examined several deep-sea drill sites from upstream of NADW, i.e., the Norwegian-Greenland Sea and the northern North Atlantic (Fig. 1), and have (re)dated some critical stratigraphic intervals. We have also assembled data from downstream of NADW to examine the global effects of the onset of NADW.

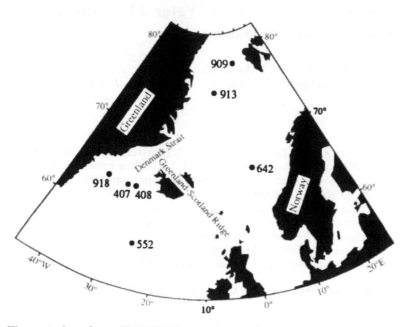

Figure 1. Locations of DSDP/ODP sites discussed in this paper.

THE NORWEGIAN-GREENLAND SEA

At ODP Site 642 in the Norwegian Sea (Fig. 1), the interval below 164 m contains no carbonate (Fig. 2). This suggests that the Norwegian Sea water was relatively stagnant and the bottom water was corrosive to carbonate. The sediment above 164 m is calcareous, with abundant nannofossils, planktonic and benthic foraminifers. This indicates the presence of NADW, which ventilates the Norwegian Sea and enhances the production and preservation of calcareous planktonic and benthic organisms. The commencement of carbonate deposition at 164 m is older than the last occurrence of the nannofossil *Coccolithus miopelagicus* (10.8 Ma) [39, 29] and the first occurrence of the foraminifer *Neogloboquadrina acostaensis* (10.7 Ma) [22, 35]. It is younger than the last occurrences of the nannofossil *Cyclicargolithus floridanus* (11.6 Ma) [39] and *Calcidiscus premacintyrei* (12.1 Ma) [39] as both species are absent above 164 m. The initiation of NADW is thus dated between 10.8 Ma and 11.6 Ma (within nannofossil zone NN7) at this site.

At ODP Sites 909 and 913 in the northern Norwegian-Greenland Sea (Fig. 1) laminated and color-banded sediments occur from ~44 Ma to ~11 Ma [26] (Fig. 2). This indicates at least interrupted presence of anoxic deep waters up to about 11 Ma. Thus, there was no significant deep water production in the Norwegian-Greenland Sea before this time. Deep water production apparently began to form at about 11 Ma, when sediment lamination terminated as a result of better ventilation in the basin. Carbonate is generally not preserved at these sites as water has been too deep (present water-depths >2500 m).

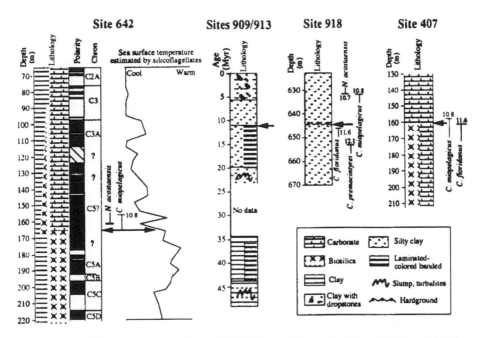

Figure 2. Stratigraphy of Sites 642 and 909/913 in the Norwegian-Greenland Sea and Sites 918 and 407 south of the Denmark Strait. Level interpreted to mark the onset of NADW is indicated by an arrow. We have (re)dated it to be older than the first occurrence of *N. acostaensis* and the last occurrence of *C. miopelagicus* (10.8 Ma) and younger than the last occurrence of *C. floridanus* (11.6 Ma). The age is best constrained at Site 918 at ~11.5 Ma. Incorporated are published data for lithology [11] and magnetostratigraphy [6] of Site 642, temperature curve based on silicoflagellates [19], Sites 909/913 [26], lithology of Site 918 [18] and Site 407 [20], and the first occurrence of *N. acostaensis* at Site 642 [34] and Site 918 [18].

NORTHERN NORTH ATLANTIC

Erosion features of ~11 Ma age can be found outside the Norwegian-Greenland Sea. At ODP Site 918 immediately south of the Denmark Strait (Fig. 1) glauconitic hardground and glaucony rip-up clasts occur at 644.0 m (Fig. 2). The hardground and rip-up clasts were apparently caused by increased bottom water currents, most likely related to the onset of NADW, when Norwegian-Greenland Sea deep water began to overflow the Denmark Strait sill. This is substantiated by benthic foraminiferal data, which show that assemblages with Norwegian Sea affinity began to occur above 644.0 m [18]. The event at 644.0 m is well constrained by a number of nannofossil and foraminiferal datums (Fig. 2); It is slightly above the last occurrence of *C. floridanus* (11.6 Ma) and can be assigned an age of 11.5 Ma.

Further evidence for the onset of Denmark Strait Overflow Water at about 11.5 Ma comes from DSDP Site 407 (Fig. 1), where there is an abrupt change from siliceous nannofossil chalk below to nannofossil chalk above 160.7 m (Fig. 2). This termination of siliceous deposition is attributed to the onset of NADW, which redistributes silica to the Indian and

Pacific oceans, as it does today. The lithologic boundary is older than the last occurrence of
C. miopelagicus (10.8 Ma) and coincident with the last occurrence of C. floridanus (11.6
Ma). An unconformity is likely to be present here, as the upper stratigraphic range of C.
floridanus appears to be truncated by this lithologic boundary and Baldauf and Barron
(1990) indicate a hiatus of about 1 m.y [1]. This is further manifestation of increased bottom
water currents associated with the onset of Denmark Strait Overflow Water.

DSDP Site 552 (Fig. 1) is currently on the path of Wyville-Thompson Ridge Overflow
Water across the eastern part of the Greenland-Scotland Ridge and may be used to monitor
the onset of this overflow water. Indeed, a major hiatus is present which separates
nannofossil zone NN7 from Paleogene sediment below (see figures 2 and 3 of Keigwin et
al., 1987) [14]. We suggest that the initiation of Norwegian Sea Overflow Water eroded the
sediment record back to the Paleogene and this event occurred around 11.5 Ma (lower part
of nannofossil zone NN7). The 11.5 Ma erosion/hiatus event identified at Sites 407, 552,
and 918 appears to correlate with the regional seismic reflector "Merlin", which was
previously identified by Mountain and Tucholke (1985) [24], who assumed a general age of
10-12 Ma.

"SILICA SWITCH"

In the tropical Indian Ocean, the deposition of biogenic silica undergoes a dramatic shift at
the middle/late Miocene boundary at about 11 Ma. Oligocene through middle Miocene
sediments contain virtually no silica whereas late Miocene through Holocene sediments
contain common silica and well preserved radiolarians [13]. This silica deposition pattern is
the same in the North Pacific but is reversed in the North Atlantic (see Site 407, Fig. 2).
This "silica switch" has long been recognized and ascribed to the formation of NADW [2],
although the age was previously not well constrained (10-15 Ma according to Keller and
Barron, 1983; Woodruff and Savin, 1989) [16, 36]. The relatively large age uncertainty was
due to the compilation of data sets from a large number of sites with vastly different data
quality in terms of age assignments and placements of the sedimentary facies change by
various authors.

CARBON ISOTOPE DATA

The onset of NADW around 11.5 Ma is also suggested by carbon isotope data [36] (Fig. 3).
Before 11.5 Ma, $\delta^{13}C$ values in the deep Atlantic are virtually the same as in the deep
Pacific, indicating that water mass did not age progressively from the Atlantic to the Pacific
as it does today. After 11.5 Ma, $\delta^{13}C$ values are consistently higher in the Atlantic than in
the Pacific. This suggests that NADW began to be active, which led to progressive
enrichment of CO_2 due to oxidation of organic matter and thus lowered $\delta^{13}C$ values as water
flows from the Atlantic to the Pacific.

IMPLICATIONS

It has been proposed that the initiation of NADW led to stronger upwelling in the Southern
Ocean, and the resulting increased moisture supply to Antarctica caused the expansion of the
Antarctic ice sheet [32]. This "snow gun" hypothesis was also favored by Prentice and
Matthews (1988) to characterize the Antarctic ice sheet history [31]. However, it is now
clear that different data sets from the Norwegian-Greenland Sea, Atlantic, Pacific and Indian
oceans are all consistent in suggesting the onset of NADW at ~11.5 Ma. This is about 3 m.y.
younger than the large middle Miocene $\delta^{18}O$ increase caused by the expansion of Antarctic

ice sheet and global cooling. Consequently, the onset of NADW cannot be the cause of the ice sheet expansion in Antarctica as previously thought [32, 17].

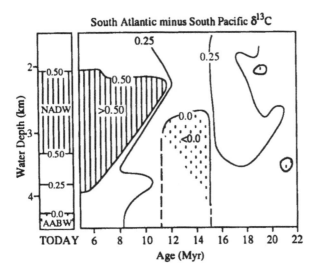

Figure 3. South Atlantic δ^{13}C minus South Pacific δ^{13}C plotted on axes of age versus backtracked water depth (simplified from Woodruff and Savin, 1989) [36]. Note that positive δ^{13}C gradients between the Atlantic and Pacific are consistently present beginning ~11.5 Ma, suggesting the onset of NADW at this time.

It has long been assumed that the subsidence of the Greenland-Scotland Ridge to a critical water depth triggered the onset of NADW [32, 5, 23], although the subsidence history of the ridge has been poorly known. The new age established here for the onset of NADW at 11.5 Ma prompts a reassessment of the general assumption. During the interval from 14 Ma to 11 Ma, sea level dropped ~220 m according to Haq et al. (1987) [12], whereas the Greenland-Scotland Ridge subsided ~130 m based on the subsidence curve of Sclater et al. (1985) [33] or the calculation of Miller et al. (1987) [23]. This suggests that the Greenland-Scotland Ridge became shallower rather than deeper during this time. If subsidence of the ridge was the triggering mechanism for the onset of NADW, NADW should have been initiated before 14 Ma and shut down when the rate of sea level drop exceeded that of subsidence of the ridge. In addition, reconstructions of the subsidence history of the Greenland-Scotland Ridge suggest that its eastern parts sank beneath sea level probably during middle Eocene time and during early to middle Miocene times in the Denmark Strait area [26]. The Wyville-Thompson Ridge in the eastern part of the Greenland-Scotland Ridge is thought to be on stretched continental crust, and this stretching probably occurred in the early Eocene and the ridge could have been deep since then [33]. Furthermore, there is an indication from Site 552 that the Wyville-Thompson Ridge Overflow Water was initiated at the same time as the Denmark Strait Overflow Water even though the sill depths of Denmark Strait and the Wyville-Thompson Ridge have been significantly different. Finally, a critical sill depth as the trigger for the onset of NADW has not been successfully modeled.

On the other hand, climate cooling may have triggered the onset of NADW, as suggested by silicoflagellate data [19] from Site 642, where the onset of NADW is within a prominent cooling in the Norwegian Sea (Fig. 2). The onset of NADW appears to coincide with one of the oxygen isotope enrichments - Mi5 [38], which is an indication of global cooling and/or ice volume increase. Global climate has cooled substantially since 15 Ma, when Antarctic ice sheet began to expand rapidly. Oxygen isotope data of planktonic foraminifers from DSDP Site 408 near the Denmark Strait show values around 1 per mil for the 11-12 Ma time interval [15]. This suggests surface water temperatures around 7°C, assuming about 1/2 of the present ice volume at that time. This means that by 11-12 Ma the surface water at Site 408 had already cooled to about its present temperature. In addition, palynological data from Iceland [25] suggest a mean annual temperature of 3-5°C by about 10 Ma. The relatively low temperatures in the northern high latitudes and thus steep equator-to-pole thermogradients by 10-12 Ma most likely increased atmospheric circulation and the flow of the North Atlantic Current into the Norwegian Sea, where the surface water was chilled below a temperature threshold and large scale thermohaline circulation began.

In any case, climate cooling must be an important factor in the onset of NADW as no strong, large thermohaline circulation could proceed without a relatively cool Norwegian-Greenland Sea. The realization of the important role that climate cooling may have played on the onset of NADW has significant implications. Large increase in temperature at high latitudes in the next few centuries as climate modelers predict may reverse the process leading to the initiation of NADW and eventually shut down NADW. If so, the North Atlantic and western Europe may become colder and a fundamentally different global climate may emerge. The potentially severe consequence of this possibility demands thorough studies of NADW in various scales and perspectives.

Acknowledgements

Supported in part by the U.S. National Science Foundation and the Petroleum Research Fund. Discussion with E. Jansen on paleoceanography of the Norwegian-Greenland Sea was very helpful. We thank W. Berger and W. Hay for reading an early version of this paper and for encouragement. W.A. Berggren and Wang Pinxian offered helpful suggestions for improvements. Samples used in this study were supplied by the international Ocean Drilling Program.

REFERENCES

1. Baldauf and J.A. Barron. Evolution of biosiliceous sedimentation patterns - Eocene through Quaternary: Paleoceanographic response to polar cooling. In: *Geological History of the Polar Oceans: Arctic versus Antarctic*. U. Bleil and J. Thiede (Eds). pp. 575-607. Kluwer Academic Pulishers, the Netherlands (1990).
2. Berger. Biogenous deep-sea sediments: Fractionation by deep-sea circulation, *Geol. Soc. Amer. Bull.* 81, 1385-1402 (1970).
3. Berger. Deep sea carbonates: Dissolution facies and age-depth constancy, *Nature* 236, 392-395 (1972).
4. Berger. Biogenous deep sea sediments: Production, preservation and interpretation. In: *Treatise on Chemical Oceanography* 5. J.P. Riley and R. Chester (Eds). pp. 265-388 (1976).
5. -L. Blanc, D. Rabussier, C. Vergnaud-Grazzini and J.-C. Duplessy. North Atlantic Deep Water formed by the later middle Miocene, *Nature* 283, 553-555 (1980).
6. Bleil. Magnetostratigraphy of Neogene and Quaternary sediment series from the Norwegian Sea: Ocean Drilling Program, Leg 104. In: *Proceedings of the Ocean Drilling Program, Scientific Results* 104. O. Eldholm, J. Thiede, E. Taylor et al (Eds). pp. 829-901. Colege Station, Texas, Ocean Drilling Program (1989).

7. Bond, W. Broecker, S. Johnsen, J. McManus, L. Labeyrie, J. Jouzel and G. Bonanl. Correlations between climate records from North Atlantic sediments and Greenland ice, *Nature* 365, 143-147 (1993).
8. Boyle and L. Keigwin. North Atlantic thermohaline circulation during the past 20,000 years, geochemical evidence, *Science* 218, 784-787 (1987).
9. Broecker and G.H. Denton. The role of ocean-atmosphere reorganizations in glacial cycles, *Geochim Cosmochim Acta* 53, 2465-2501 (1989).
10. Cande and D.V. Kent. Revised calibration of the geomagnetic polarity timescale for the Late Cretaceous and Cenozoic, *Jour. Geophys. Res.* 100, 6093-6095 (1995).
11. Eldholm, J. Thiede, E. Talyor et al. *Proceedings of the Ocean Drilling Program, Initial Results* 104. Colege Station, Texas, Ocean Drilling Program (1987).
12. Haq, J. Hardenbol and P.R. Vail. Chronology of fluctuating sea levels since the Triassic, *Science* 235, 1156-1167 (1987).
13. Johnson Radiolarian biostratigraphy in the centra Indian Ocean, Leg 115. In: *Proceedings of the Ocean Drilling Program, Scientific Results* 115. R.A. Ducan, J. Backman, L.C. Peterson et al (Eds). pp. 395-409. College Sation, Texas, Ocean Drilling Program (1990).
14. Keigwin, M.-P. Aubry and D.V. Kent. North Atlantic late Miocene stable-isotope stratigraphy, biostratigraphy, and magnetostratigraphy. In: *Initial Reports of the Deep Sea Drilling Project* 94. W.F. Ruddiman, R.B. Kidd, E. Thomas et al (Eds). pp. 935-963. Washington, D.C., U.S. Government Printing Office (1987).
15. Keigwin, W.B. Curry, S.J. Lehman and S. Johnsen. The role of the deep ocean in north atlantic climate change between 70 and 130 kyr ago, *Nature* 371, 323-326 (1994).
16. Keller and J.A. Barron. Paleoceanographic implications of Miocene deep-sea hiatuses, *Geol. Soc. Amer. Bull.* 94, 590-613 (1983).
17. Kennett. *Marine Geology*. Englewood Cliffs, New Jersey, Prentice-Hall Inc (1982).
18. Larsen, A.D. Saunder, P.D. Clift et al. *Proceedings of the Ocean Drilling Program, Initial Reports* 152. College Station, Texas, Ocean Drilling Program (1994).
19. Locker and E. Martini. Cenozoic silicoflagellates, ebridians, and actiniscidians from the Voooring Plateau (ODP Leg 104). In: *Proceedings of the Ocean Drilling Program, Scientific Results* 104. O. Eldholm, J. Thiede, E. Taylor et al (Eds). pp. 543-585. Colege Station, Texas, Ocean Drilling Program (1989).
20. Luyendyk, J.R. Cann et al. *Initial Reports of the Deep Sea Drilling Project* 49. Washington D.C., U.S. Government Printing Office (1979).
21. Miller and R.G. Fairbanks. Oligocene to Miocene carbon isotope cycles and abyssal circulation changes. In: *The Carbon Cycle and Atmospheric CO$_2$: Natural Variations Archean to Present*. E.T. Sunquist and W.S. Broecker (Eds). pp. 459-468. Washington D.C., American Geophysical Union (1985).
22. Miller, M.-P. Aubry, J. Khan, A.J. Melillo, D.V. Kent and W.A. Berggren. Oligocene-Miocene biostratigraphy, magnetostratigraphy and isotopic stratigraphy of the western North Atlantic, *Geology* 13, 257-261 (1985).
23. Miller, R.G. Fairbanks and E. Thomas. Benthic foraminiferal carbon isotopic records and the development of abyssal circulation in the eastern North Atlantic, Initial Reports of the Deep Sea Drilling Project, Volume 94. In: *Initial Reports of the Deep Sea Drilling Project* 94. W.F. Ruddiman, R.B. Kidd, E. Thomas et al (Eds). pp. 981-996. Washington, D.C., U.S. Government Printing Office (1987).
24. Mountain and B.E. Tucholke. Mesozoic and Cenozoic geology of the U.S. Atlantic continental Slope and Rise. In: *Geologic Evolution of the United States Atlantic Margin*. C.W. Poag (Ed). pp. 293-341. Van Nostrand Reinhold, NY (1985).
25. Mudie and J. Helgason. Palynological evidence for Miocene climatic cooling in eastern Iceland about 9.8 Myr ago, *Nature* 303, 689-692 (1983).
26. Myhre, J. Thiede, J.V. Firth et al. *Proceedings of the Ocean Driling Program, Initial Reports* 151. College Station, TX, Ocean Drilling Program (1995).

27. Oppo and R.G. Fairbanks. Variability in the deep and intermediate water circulation of the Atlantic Ocean during the past 25,000 years: Northern Hemisphere modulation of the Southern Ocean, *Ear. Planet. Sci. Lett.* **86**, 1-15 (1987).

28. Oppo and S.J. Lehman. Subordital timescale variability of North Atlantic Deep Water during the past 200,000 years, *Paleoceanography* **10**, 901-910 (1995).

29. Peleo-Alampay. Morphometric and Biochronologic study of *Coccolithus miopelagicus*, the 6th International Nannoplankton Association Conference, Copenhagen, Programme and Abstracts, 93 (1995).

30. Pillar and L. Labeyrie. Role of the thermohaline circulation in the abrupt warming after Heinrich events, *Nature* **372**, 162-164 (1994).

31. Prentice and R.K. Matthews. Cenozoic ice-volume history: Development of a composite oxygen isotope record, *Geology* **17**, 963-966 (1988).

32. Schnitker. North Atlantic oceanography as possible cause of Antarctic glaciation and eutrophication, *Nature* **284**, 615-616 (1980).

33. Sclater, L. Meinke, A. Bennett and C. Murphy. The depth of the ocean through the Neogene. In: *The Miocene Ocean: Paleoceanography and Biogeography.* J.P. Kennnett (Ed). pp. 1-10. Geol. Soc. Amer. Mem. 163 (1985).

34. Spiegler and E. Jansen. Planktonic foraminifer biostratigraphy of Norwegian Sea sediments: ODP Leg 104. In: *Proceedings of the Ocean Drilling Program, Scientific Results* 104. O. Eldholm, J. Thiede, E. Taylor et al (Eds). pp. 681-696. Colege Station, Texas, Ocean Drilling Program (1989).

35. Weaver and B.M. Clement. Magnetobiostratigraphy of planktonic foraminiferal datums: Deep Sea Drilling Project Leg 94, North Atlantic. In: *Initial Reports of the Deep Sea Drilling Project* 94. W.F. Ruddiman, R.B. Kidd, E. Thomas et al (Eds). pp. 815-829. Washington, D.C., U.S. Government Printing Office (1987).

36. Woodruff and S.M. Savin. Miocene deepwater oceanography, *Paleoceanography* **4**, 87-140 (1989).

37. Wright and K.G. Miller. Control of North Atlantic Deep Water circulation by the Greenland-Scotland Ridge, *Paleoceanography* **11**, 157-170 (1996).

38. Wright, K.G. Miller and R.G. Fairbanks. Early and middle Miocene stable isotopes: Implications for deepwater circulation and climate, *Paleoceanography* **7**, 357-389 (1992).

39. Young, J.-A. Flores and W. Wei. A summary chart of Neogene nannofossils magnetobiostratigraphy, *Jour. Nannoplankton Res.* **16**, 21-27 (1994).

Proc. 30ᵗʰ Int'l. Geol. Congr., Vol. 13, pp. 65-86
Wang & Berggren (Eds)
©VSP 1997

West Pacific Marginal Seas during Last Glacial Maximum: Amplification of Environmental Signals and Its Impact on Monsoon Climate*

PINXIAN WANG[1], M. BRADSHAW[2], S.S.GANZEI[3], S.TSUKAWAKI[4], K.BIN HASSAN[5], W.S.HANTORO[6], S. POOBRASERT[7], R.BURNE[2], QUANHONG ZHAO[1], H.KAGAMI[8]

[1] *Department of Marine Geology, Tongji University, Shanghai 200092, China*
[2] *Australian Geological Survey Organization, GPO Box 378, Canberra 2601, Australia*
[3] *Pacific Institute of Geography, Russian Academy of Science, Vladivostok 690041,Russia*
[4] *Department of Geology, Kanazawa University, Kanazawa, 920-11, Japan*
[5] *Geological Survey of Malaysia, PO Box 11110, 50736 Kuala Lumpur, Malaysia*
[6] *Centre for Geotechnology, LIPI, Bandung 40135, Indonesia*
[7] *Department of Geology, Chulalongkorn University, Bangkok, Thailand*
[8] *Institute of Geology, Faculty of Science, Josai University, Nakado, 350-02 Japan*

Abstract

A paleogeographic map has been compiled for the time slice of the last glacial maximum (15-20 kaBP) covering the Western Pacific region with emphasis laid on the marginal seas. This UNESCO/IOC map (1:20,000,000) is based on paleogeographic and paleoenvironmental data from 779 offshore and onshore sites, and about five hundred publications have been collected for this purpose. As seen from the Paleogeographic Map, the emergence of extensive continental shelves was the most outstanding geographic feature of the last glacial maximum in the West Pacific region. The sea-level induced environmental signal has been amplified in the marginal seas, giving rise to drastic changes in sea areas and configurations, and to reorganization of sea water circulation in seas of enclosed basin type. Since most of the Western Pacific marginal seas are influenced by monsoon circulation and some of these are located within the Western Pacific Warm Pool, the glacial geographic changes have produced a profound impact on regional and global climate. For example, the decrease of sea area and sea surface temperature (SST) in the marginal seas was one of factors responsible for the enhanced aridity of inland China during the glaciation. Glacial intensification of the winter monsoon and increased seasonality of SST in marginal seas might explain, at least partly, the apparent discrepancy between the tropical paleotemperature estimations based on terrestrial and open-ocean records in this region.

Keywords: paleogeography, paleoceanography, sea-level changes, Western Pacific, marginal seas, last glacial maximum, monsoon

---

* All correspondence should be addressed to the first author

INTRODUCTION

The Western Pacific region, with its numerous sea ways and vast coastal lowlands and shelves, has been very sensitive to sea-level changes.   The late Quaternary glacial cycles have witnessed drastic environmental changes in this region related to eustatic fluctuations.Therefore, paleogeographic reconstruction in the Western Pacific offers special scientific and social-economic interests in environmental prediction and coastal management.

An international working group was set up by the Intergovernmental Oceanographic Commission (IOC) / UNESCO in 1990 to produce a series of late Quaternary Paleogeographic Mapping Working Group compiled the first maps for the last glacial maximum (LGM, 15-20 kaBP). The paleogeographic map (1:2,000,000), accompanied by two data maps at scale 1:1,000,000 for the Northern and Southern Hemispheres and by a volume of explanatory notes, was published by UNESCO / IOC (1995) [30](copies are available upon request). Paleo-coastlines, sea surface temperature, sediment types, sea-ice limits, some geomorphological and other features are shown on the paleogeographic map. Although the maps cover a broad area to include the Western Pacific, from 90°E to 170°W and from 65°N to 65°S, the emphasis has been laid on the marginal seas. Seven countries of the Western Pacific region participated in compilation of the map: Australia, China, Indonesia, Japan, Malaysia, Russia and Thailand, but the main components of the map come from three source maps provided by Russia (A in Fig.1), China (B in Fig.1) and Australia (C in Fig.1). The Working Group was led by Pinxian Wang and Marita Bradshaw, with other co-authors of this paper being its members.
A wealth of data from the region has been pooled together to produce the maps. A total of 779 offshore and onshore sites with paleogeographic information of the LGM were collected and utilized for compilation of the maps, of those 172 sites have been radiocarbon dated, 157 sites provided with oxygen-isotope dates, 61 sites with tephrochronological or other datings, 383 sites analysed for microfossils, 53 for pollen, and 86 sites dated on the basis of carbonate stratigraphy. Meanwhile, nearly five hundred of publications with various paleogeographic information of the LGM in the Western Pacific region, written in different languages in different countries, have been collected, and the bibliography is provided in the explanatory volume for the reader's reference.

The Paleogeographic Map and collected data have revealed the variety of marginal seas, their response to glacial cycles, and their climate impact on the entire region. The present paper is a discussion on the basis of the maps and the supporting data to demonstrate geographic changes in the marginal seas at the LGM and environmental consequences caused by the glacial sea-level lowering. We show that the role of marginal seas has been underestimated in paleoenvironmental reconstruction and climate modelling, and the special features of monsoon climate should not be ignored when environmental changes in the Western Pacific are considered.

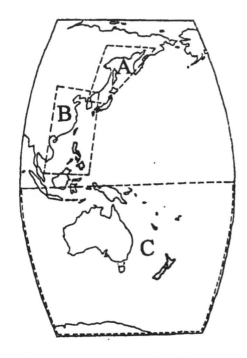

Figure 1. Three main source maps for the WESTPAC Paleogeographic Map, contributed by individual countries: A. Russia; B. China; C. Australia.

EMERGENCE OF SHELVES

The LGM was distinguished by the development of enormous ice sheets. The last glacial Laurentide ice sheet in North America, for instance, was similar in area to the present Antarctic ice sheet, and the Scandinavian ice sheet then covered all the Barents Sea and the Kara Sea, but no continental ice cap was developed in northeast Asia [7]. Instead, the emergence of vast shelf areas of the marginal seas was the most prominent geographic change in the Western Pacific region at the LGM.

The LGM geography of the marginal seas can be approximated by today's configuration with a lowered sea level. The estimation of the last glacial low stand of sea level, however, varies from author to author at least from -85m to -135m [6]. Although sea level changes varied considerably with location, the West Pacific marginal seas lie in one and the same sea level zone (zone III, Clark & Lingle, 1976 [5]) and should have a similar LGM low stand. On the other hand, the area in question is too heterogeneous in tectonic settings and some tectonically active coasts may display very high rates of uplift or subsidence. For example, the eastern part of Indonesia and Malaysia uplifted 81-162m since 18,000 aBP, whereas in Semporne, NE Borneo, the paleo-shoreline marker dated 19,030 + 450 aBP occurs at + 2m above the present sea level [28]. Fortunately, tectonically active sea areas, as a rule, slope steeply down to the deep basin and the magnitude of possible errors in low stand estimates would have in most cases little effect on our discussion.

P. Wang et al.

The geographic response to sea-level change of the marginal seas varies greatly depending on their morphology. Table 1 shows morphological features for twelve of the Western Pacific marginal seas included in our maps. The Timor Sea, Arafura Sea and Gulf of Carpentaria are often ascribed to the Indian Ocean, they are included here in view of their relevance to the environmental changes of the region. Tectonically, some of the listed seas are not marginal basins in the proper sense, such as the Yellow Sea and Bohai Gulf which are parts of the inner shelf of the East China Sea, and the Gulf of the Carpentaria which is an extension of the Arafura Sea shelf. As seen from Table 1, many of the seas are characterized by extensive areas of continental shelves which were exposed subaerially during the LGM (Fig. 2, black areas).

Among the shelves emerged during the LGM (black areas in Fig. 2), three are most extensive: 1) *East China Sea Shelf* with a total area of 850,000 sq.km; 2) *Sunda*

Table 1. Morphological features of the Western Pacific marginal seas[*]

Type	Marginal sea	Area[*] (km²)	Water depth (m) aver.	Water depth (m) max.	Proportion of shallow sea area (<200 m, %)	Basin depth B (m)	Sill depth S (m)	S/B
Shallow bank type	Yellow Sea, Bohai Gulf	457,000	40	140	100%	140	140	1
	Java Sea	433,000	46	1,272	90.4%	1,720	1,720	1
	Arafura Sea	650,000		3,650	90%	3,650	1,480	0.41
	Gulf of Carpentaria	511,000	39	69	100%	69	~50	0.72
Open basin type	Okhotsk Sea	1,583,000	777	3,374	41.2%	3,374	~2000	0.59
	East China Sea	770,000	370	2,719	75.6%	2,719	>2000	0.74
	Banda Sea	470,000	2,737	7,440	8.3%	7,440	3,130	0.42
	Timor Sea	450,000		3,310	ca 80%	3,310	1,800	0.54
Enclosed basin type	Japan Sea	1,070,000	1361	4,049	26.3%	4,049	130	0.03
	South China Sea	3,500,000	1,212	5,377	52.4%	5,377	2,600	0.48
	Sulu Sea	260,000	1,570	5,580	34.3%	5,580	420	0.08
	Celebes Sea	280,000	3,364	6,200	10.8%	6,200	1,400	0.23

[*] Data of morphological features from R. W. Fairbridge, 1966 ; S. G. Gorshkov, 1974; Torgersen et al., 1983 [29]. Due to the miscellaneous data sources the figures in the table may not be consistent in terms of sea area, etc.

Shelf or *Great Asian Bank*, including the southern part of the South China Sea with the Gulf of Thailand, and the Java Sea, with a total area of 1,800,000 sq.km; 3) *Sahul Shelf* or *Great Australian Bank*, including the Timor Sea shelf of Sahul shelf *s.s.*, the Arafura shelf and the Gulf of Carpentaria, with a total area of 1,230,000 sq.km. Thus, the 3 major shelves along sum to ca. 3,900,000 sq.km, equal to the entire Indian Subcontinent with India, Pakistan and Bangladesh taken together [35].

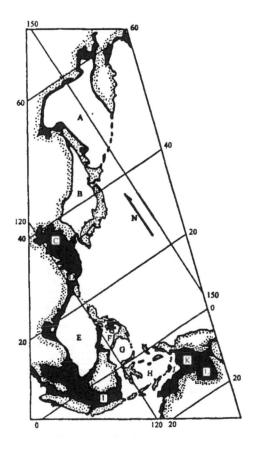

Figure 2. Western Pacific marginal seas at the last glacial maximum. Black area denotes emerged shelves. A. Okhotsk Sea; B. Sea of Japan; C. Yellow Sea & Bohai Gulf; D. East China Sea; E. South China Sea; F. Sulu Sea; G. Celebes Sea; H. Banda Sea; I. Java Sea; J. Timor Sea; K. Arafura Sea; L. Gulf of Carpentaria.

It is worthy to note that the latter two shelves (Sunda and Sahul) are located in the warm water pool of the Western Pacific and eastern Indian Oceans bounded approximately by the 28°C SST isotherm (Fig.3) [42]. The modern *warm pool* is the region with the highest open ocean surface temperature, and the global atmosphere is extremely sensitive to the SST changes in the warm pool. The glacial exposure of shelves and a decrease of SST in the deeper water parts of marginal seas (see below) within the warm pool must have had serious climate consequences.

An example is the *South China Sea*. As found by Chinese meteorologists on the basis of their numerical modeling, the South China Sea, the Bengal Bay and the West Pacific are heating centres in summer for East Asian monsoon [2], and the

P. Wang et al.

South China Sea is the main source area of moisture for the summer monsoon rain
in China [3]. Since the evaporation from sea is much higher than from land (Table
2) , the glacial reduction of sea area caused by sea-level lowering must have led to a
decrease of vapour supply from the sea, and, hence, to a *decline of monsoon
precipitation* and an increase of aridity in the China hinterland. According to
various authors, the world average difference between evaporation rates from sea and
land is estimated as 33.8 cm/a [11] or 50 cm/a [8]. The shelf areas of the South
China Sea amount ca. 1,800,000 km², and their exposure would result in a reduction
of the annual evaporation from the South China Sea by 600 or 900 km³, when the
above two world average estimations are used. This roughly equals 1/10 or 1/7 of the
total annual precipitation in all of China (6190 km³).

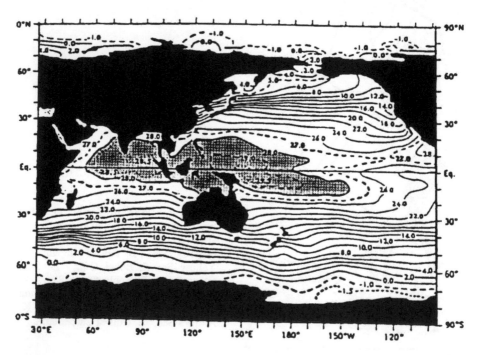

Figure 3. Modern Western Pacific Warm Pool (Modified from Yan et al., 1992 [42])

Table 2.   Different average evaporation rates between land and sea   (data from Gross,
1987[8])

area	precipitation, cm/a	evaporation, cm/a
global (aver.)	82	82
ocean (aver.)	88	97
continent (aver.)	67	47
difference	21	50

These rough estimations are only to show qualitatively the importance of marginal seas in glacial climate changes, and have no pretensions to quantitative significance. Nevertheless, it is clear that the shelf emergence of the South China Sea is one of factors responsible for the glacial intensification of aridity in China. In fact, the expansion of loess deposition area across the Yangtze River, the southeast shift of vegetation zones and the complete disappearance of the monsoon rain forests in South China, are all indicative of increased aridity and enhanced eolian processes at the LGM [38].

CHANGES IN SURFACE CIRCULATION AND TEMPERATURE

When the 12 seas are considered in the aspect of their response to low sea-level stand at glaciation, three types can be distinguished (Table 1; Fig.4):

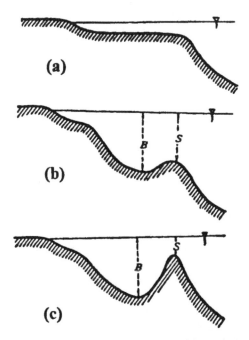

Figure 4. Schematic profiles of marginal seas. (a) Shallow bank type; (b) Open basin type; (c) Enclosed basin type. B—Basin depth; S—Sill depth.

1) *Shallow bank type* (Fig.4a) including the Yellow Sea with Bohai Gulf, the Java Sea, the Arafura Sea with the Gulf of Carpentaria. These are not marginal basins in the tectonic sense but shelf extensions of marginal seas. The shallow water bottom of these seas was entirely or mostly exposed subaerially during the LGM and covered by a network of river channels as reported from the Java Sea [31] and Yellow Sea [21], or occupied by a lake as in the case of the Gulf of Carpentaria [29].

2) *Open basin type* (Fig.4b) represented by the Sea of Okhotsk, the East China Sea and the Timor Sea where the sill depth (S in Fig.4) is more or less close to the basin depth (B in Fig.4) (S/B > 0.5). These sea basins are well connected with the open ocean, even at the lowered sea level stand of the LGM.

3)   *Enclosed basin type* (Fig.4c) such as the Sea of Japan, South China Sea, and
Sulu Sea. As the sill depth is much shallower than the basin depth (S/B<0.5), their
connection with the open ocean was restricted during the LGM sea-level fall.
Depending on the width of the connecting strait and the depth of the sill, the sea
basins displayed various extents of isolation and reorganization of the current system
at the low sea-level stand, representing the most interesting type of marginal seas in
paleoceanography.

The three types of marginal seas responded to the glacial low sea-level stand in
different ways: the seas of shallow bank type simply emerged at the LGM; the seas
of enclosed basin type displayed most complicated changes with a reorganization of
sea water circulation resulting from the restricted connection with the open ocean at
low sea level stand; less changes happened with the seas of open basin type which
retained their free connection with ocean with lowered sea level. Aside from the ice-
covered parts of the Sea of Okhotsk and the Sea of Japan, marginal seas of open
basin type underwent no fundamental changes at the LGM. Special discussions are
needed on the glacial circulation in the seas of enclosed basin type and in the West
Pacific as a whole.
The glacial surface circulation pattern in the West Pacific differs from the modern
one. The polar front in the North Pacific shifted southward about 10 degrees of
latitude during the last glaciation [26], and the westerly winds in the Southwest
Pacific shifted significantly to the South [24]. Using four groups of plankton
microfossils, Chinzei et al.(1987)[4] were able to demonstrate the postglacial
northward migration of the Kuroshio Front off the southeastern coast of Honshu,
implying that the ice-age location of the Kuroshio Front was to the south of Honshu.
At the same time, much more essential changes of circulation took place in enclosed
basin type seas such as the Sea of Japan.

The modern *Sea of Japan* is distinguished from all the other marginal seas in the
West Pacific by its shallow sills. Of the four connecting straits only two were deep
enough to maintain water exchanges with the Pacific (Tsugaru Strait, sill
depth130m) and the East China Sea (Tsushima Strait, sill depth 130m) during the
LGM. As the sill depths are very close to that of the LGM low sea-level stand, the
Sea of Japan was almost isolated from the ocean. The heavily restricted connection
with the ocean water had led to stratification of the water column, with diluted water
at the surface and low oxygen content of the bottom water, as indicated by the thick
layer of dark olive gray, thinly laminated clay which is characterized by the absence
of burrowed structure and benthic foraminifers. The very light oxygen-isotopic
values of planktonic foraminifers, dated to 27-20 kaBP [20], also support this
interpretation.

Less drastic changes have been recorded in the Sulu Sea and the South China Sea,
which are, however, more significant for the hydrothermal budget because of their
position at lower latitudes. The *Sulu Sea* is also enclosed by shallow sills except for
two deeper channels at its northeast and east: one off Panay and another between
Nigros and Mindanao with sill depths of 420 and 250m, respectively. It was
expected that the shallow sills would have led to restricted water exchange with the
ocean and, again, to low oxygen content in the bottom water. The ODP pre-site
survey, however, shows that the postglacial and glacial deposits consist of olive gray

foraminiferal-nannofossil marl with intensive bioturbation, suggesting well oxygenated bottom conditions probably related to turbidite action[32]. The increased abundance of *Neogloboquadrina dutertrei* and light oxygen isotope values of planktonic foraminifers indicate reduced salinity of the surface water at LGM, and the high percentage of *Bolivina robusta* infauna from mid-water depth and the benthic carbon isotope data suggest an expansion of oxygen minimum zone in its northwest part [15].

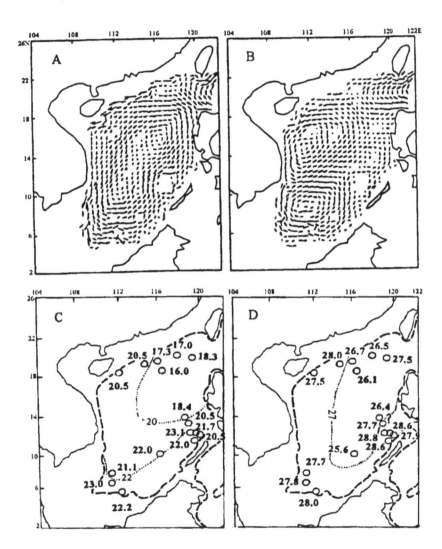

Figure 5. Comparison between numerically simulated surface circulation and transfer function temperature pattern in the South China Sea at the LGM. Surface circulation: A. winter, B. summer; Surface temperature: C. winter, D. summer. [37]

As to the *South China Sea*, the only connection to the Pacific during the LGM was the Bashi Strait; but its great sill depth (about 2600m) would not hamper the water exchanges with Pacific or lead to any stratification of its water column. Rather, the glacial cut-off of its southern connections to the Indian Ocean must have given rise to a fundamental reorganization of surface circulation in the South China Sea (SCS). Our numerical modelling has shown that the modern trans-basinal circulation driven by the Asian monsoon was replaced at the LGM by a semi-enclosed pattern, with a clockwise circulation for summer and a counter-clockwise circulation in winter (Fig. 5 A, B [37]). Together with the southern shift of the north Pacific polar front and the strengthened north winds of the winter monsoon, the glacial circulation pattern in the SCS resulted in abnormally low winter SSTs as compared with those in the open ocean at similar latitudes, in an enhanced north-south gradient of winter SST and an increased seasonality (Fig. 5 C, D [39]).

CONTRAST IN SEASONALITY

One of the major findings by CLIMAP was that the "ice-age ocean was strikingly similar to the present ocean in at least one respect: large areas of the tropics and subtropics within all oceans had sea-surface temperatures as warm as, or slightly warmer than, today" (CLIMAP, 1981[6], p.9). This can be demonstrated with the *sea surface temperature* (SST) reconstructions of the Western Pacific region based on micropaleon-tological census using the transfer function technique (Fig. 6; [18]). Recent studies by Thunell et al. (1994)[27] on the tropical Western Pacific paleo-SST at the LGM using the modern analogue technique (MAT) also "indicate that tropical SSTs differed by less than 2°C from present" (Fig. 7). This is supported by the oxygen isotopic data of the Western Pacific and used to indicate the stability of the Western Pacific Warm Pool during the glacial cycles [27].

However, the CLIMAP paleo-SST reconstructions in this region were based on very limited site studies, and the MAT paleo-SST gives only annual average. Since the Western Pacific marginal seas offer better sediment records than the open ocean and display strong influence of monsoon circulation, it is essential to consider the seasonal SSTs in the marginal seas at the LGM. During recent years, winter and summer paleotemperature estimations for the LGM have been obtained from many cores in the South China Sea, based on planktonic foraminiferal census data using Transfer Function FP-12 [25]. Wang and Wang (1989, 1990) [33,34] have found that glacial/interglacial fluctuation in SST and the LGM *seasonality* are much higher in the northern part of the SCS than the adjacent open Pacific, reflecting the amplification of the climatic signal in the marginal sea. Later, this was confirmed by data from the southern part of the SCS [17], as well as by paleotemperature estimations based on organic geochemical proxy U_k^{37} and the modern analogue technique of paleo-SST estimations [10]. The streng-thened winter SST contrast between the marginal seas and the open ocean and the enhanced seasonality at the LGM have been also found in the East China Sea [41, 12] and the Sulu Sea [17].

Available paleo-SST data from the South China Sea, East China Sea, Sulu Sea and the adjacent west Pacific at the LGM are summarized in Figs. 8 and 9. The paleo-SST estimations are based on planktonic foraminiferal census using Transfer Function FP12-E developed for the Western Pacific [25]. As seen from the figures,

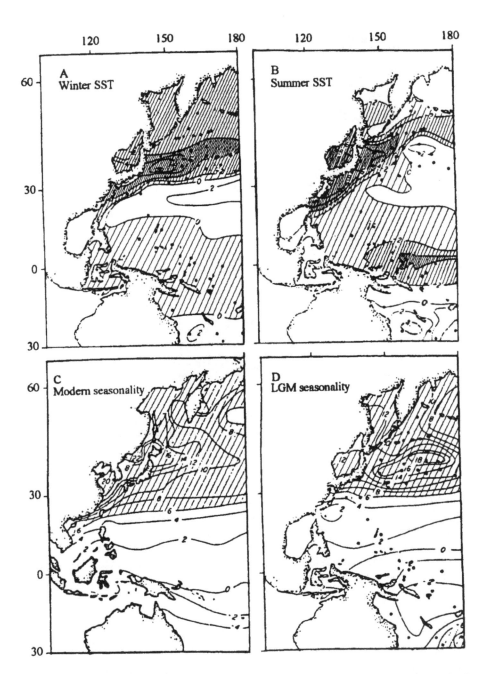

Figure 6. Sea surface temperature reconstructions for the Western Pacific region at the last glacial maximum by CLIMAP. A. Winter SST difference between LGM (about 18 kaBP) and present; B. Summer SST difference between LGM and present; C. Modern seasonality (difference between summer and winter SST); D. LGM seasonality . (modified from Moore et al., 1980[18])

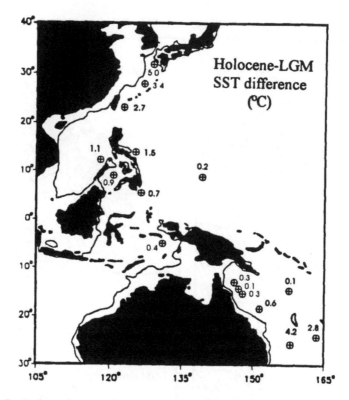

Figure 7. Estimated sea surface temperature difference between LGM and present in the Western Pacific region (annual average SST based on modern analogue technique) (After Thunell et al., 1994 [27]).

there was a great seasonal difference in SST patterns for the low latitude Western Pacific at the LGM: in winter the SST was much cooler in the marginal seas than in the open Pacific (Fig. 8), whereas in summer the SST was similar in the marginal seas and ocean (Fig. 9), resulting in a much more intensive seasonality at the LGM in the marginal seas (Fig. 10). In fact, the winter SST at the LGM was at least 3-4°C lower in the South China Sea and Sulu Sea than in the open Pacific (Fig. 8); the same is true for the seasonality (Fig. 10) and the glacial/interglacial contrast in oxygen isotope values of planktonic foraminifers (Fig. 11).

MONSOON CLIMATE

Since moisture in the East Asian continent, and China in particular, is mainly supplied by southern (boreal summer) monsoon, while precipitation in northern Australia and the islands between Australia and the Asian continent is provided mainly by northern (boreal winter) monsoon, the seasonal difference in SST pattern in the marginal seas should have different climate impact on the two areas.

As shown above, the glacial reduction of sea area should decrease the evaporation rate from the sea surface. Given wind direction and intensity unchanged, the *summer monsoon* would have brought less precipitation to the East Asian land during the LGM because of the cooler SST. The decrease in SST and in evaporation then must have intensified the aridity inland.

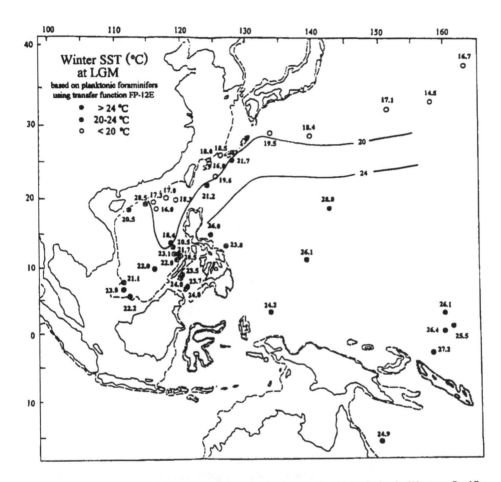

Figure 8. Winter sea surface temperature in the low and middle latitude Western Pacific and marginal seas at the last glacial maximum. The paleo-SST estimations are based on census of planktonic foraminifers using Transfer Function FP12-E. [36]

On the other hand, the *winter monsoon* in the Northern Hemisphere becomes the summer monsoon in the Southern Hemisphere when across the Equator. This is the source of precipitation in northern Australia. The north (northeast) winds of the winter monsoon also provide the main part of annual precipitation for islands around the southern South China Sea, including Kalimantan (Borneo), Sumatra, Java, Philippines, Malaysia, the Vietnam coast, etc. The mountainous relief leads to rainfall from the moisture-full airmass crossing the equatorial South China Sea from the north [13]. Again, a close link exists between the SST variations and winter monsoon rain there (e.g., Lim and Tuen, 1991[14]).

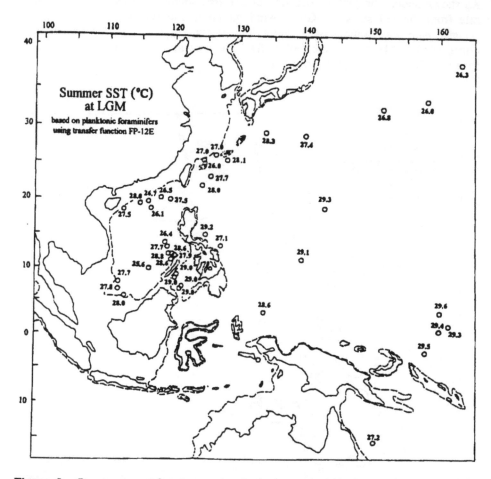

Figure 9. Summer sea surface temperature in the low and middle latitude Western Pacific and marginal seas at the last glacial maximum. The paleo-SST estimations are based on census of planktonic foraminifers using Transfer Function FP12-E. [36]

In the LGM, the strengthened winter Asian monsoon, together with the low anomaly of winter SST in the low latitude marginal seas of the Western Pacific, would give rise to a cool and humid climate in the region discussed above. This is confirmed by the terre-strial paleoclimate records there: In New Guinea the snowline at the LGM was about 1000 m lower than it is now [40], and the vegetation zones in Sumatra and Java shifted several hundreds meters downslope at the LGM [23], suggesting a 6 to 8°C decrease in air temperature in the tropical mountains. The scientific community has since long been puzzled by the discrepancy between terrestrial and marine climate records: according to CLIMAP, the SSTs in the tropics at the LGM were almost unchanged compared with the present values, whereas the lower snowline and vegetation changes require much more cooling (Fig. 12). It was suggested that the clue to the tropical climate enigma may lie in one of the two hypothesis[1] :

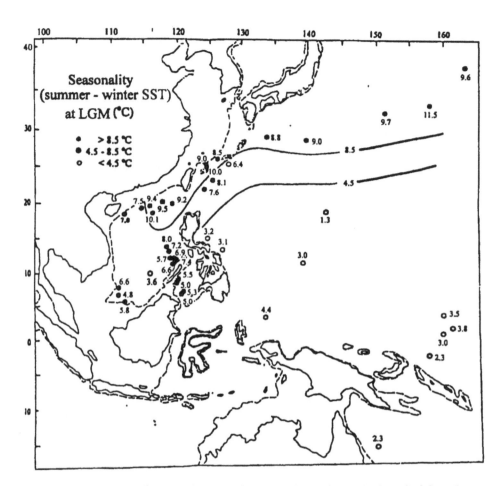

Figure 10. Seasonality in SST (summer SST minus winter SST) at the last glacial maximum in the low and middle latitude part of the Western Pacific. [36]

1)   *Atmospheric lapse rate.* The lower snowline and vegetation-zone boundaries can be partly explained, if the lapse rates (the changes of air temperature with hight) in mountain areas had been steeper in the glacial time than today [40, 22, 27]. However, the recent studies of noble gas content have shown unchanged lapse rates at glacial times, at least for southwest North America [1] and the idea was practically rejected.

2)   *SST estimation.* Recent studies on Barbados corals have revealed that tropical SSTs at the LGM were 5°C colder than at present [9], suggesting that the CLIMAP's statistically based SSTs from microfossils underestimated the glacial/interglacial constrast in the tropics.   Cooler SSTs in tropical oceans would fit the terrestrial records for the glaciation.   However, the CLIMAP paleo-SSTs are derived from multiple proxy sources from each of the tropical oceans, based on four groups of plankton microfossils and are supported by the oxygen-isotope data. Hence, these estimates can not be easily dismissed.   It is difficult to say which estimations are

reliable, as much more work in ecology and physiology is required to understand the nature of the SST proxies based on different groups of organisms [1].

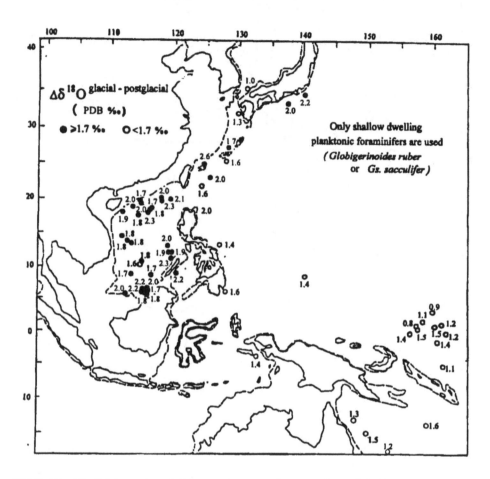

Figure 11. Glacial / postglacial difference in oxygen isotope values of shallow-dwelling planktonic foraminifers in the low and middle latitude part of the Western Pacific. [36]

The above discussed winter SST pattern in the Western Pacific related to the Asian monsoon may offer another explanation for the discrepancy between sea and land for the region. The lower winter SSTs in the marginal seas than in the open ocean at the LGM must be real, as the estimations are derived from the same group of organisms and by the same technique. The enhanced winter (in the Northern Hemisphere) monsoon at the glacial had resulted in significant cooling of winter SSTs in the marginal seas, which in turn gave rise to low atmospheric temperatures in New Guinea, north Australia, Indonesia etc., although the open ocean SSTs then were close to those of the present. Meanwhile, the intensified northern winds of the boreal winter monsoon could bring moisture to those areas, compensating at least partly the reduction in evaporation caused by the emergence of shelves and the decrease of SST there.

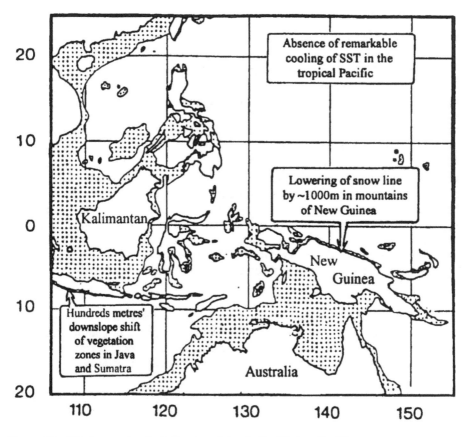

Figure 12. The ice-age tropical climate enigma in the Western Pacific

Thus, the *intensified winter monsoon in the marginal seas* may be the clue to the tropical climate enigma. The apparent contradictions between the land and sea proxies of glacial climate arise when merely open ocean conditions are taken into consideration. However, the role of marginal seas is essential when climate evolution of the neighbouring land or islands is discussed; and the seasonality is crucial when climate changes in a monsoon region are considered. Seasonality may vary greatly with the annual SST almost unchanged. An example is Site GGC-9 from the South China Sea (11°37'N, 118°37'E). The coretop-glacial departure of annual average SST is only 1.1°C [27], while that of winter SST ("Glacial-Holocene anomaly") reaches 7.3°C [17]. The annual averages of SST, therefore, obscure the specific characters of the monsoon climate and those of the marginal seas. It would be very difficult to find an explanation for the glacial lowering of snowline in New Guinea, when both specific features of this region, monsoon and marginal sea, are ignored.

CONCLUSIONS

1.   The emergence and submergence of the extensive continental shelves are the most outstanding geographic features in late Quaternary glacial cycles in the West Pacific region. The sea-level induced environmental signal has been amplified in the marginal seas, giving rise to drastic changes in sea areas and configurations, and to reorganization of sea water circulations in seas of enclosed basin type.

2.   The glacial intensification of the winter monsoon and reorganization of circulation has resulted in increased seasonality of SST and enhanced glacial /interglacial contrast of winter SST in the marginal seas, as compared with the open ocean at similar latitudes. A climatic impact of this increased seasonality of the marginal seas was reduced evaporation which intensified glacial aridity in the East Asia continent and in China, in particular. On the other hand, the low winter SST together with the strong northern monsoon wind led to a lowered snowline in New Guinea and lowered vegetation boundaries in Sumatra and Java at the LGM. The winter monsoon and the marginal seas may be responsible for the apparent discrepancy between the tropical paleotemperature estimations based on terrestrial and open ocean records.

3.   The Western Pacific Warm Pool must have been reduced in size during the last glacial maximum. This was caused by the glacial emergence of continental shelves and the decrease of SST in the marginal seas within the modern warm pool; and by the northward migration of the Tasman Front in the Southern Hemisphere [16].

4.   The role of marginal seas should not be disregarded or under-estimated in climate modelling. Particular attention should be paid to the seas of enclosed basin type with relatively narrow and shallow connection to the ocean. Those are very sensitive even to minor sea level fluctuations. Their response to deglaciation processes, for instance, should be studied in detail with high-resolution stratigraphy in order to reveal the mechanisms of environmental change in the West Pacific region, where no extensive ice cap existed at the LGM.

Acknowledgements

This work derives from the WESTPAC Paleogeographic Mapping Project of the UNESCO/ IOC which was initiated by Dr. P.Cook. Dr. G.Kullenberg, Dr. Jilan Su and Dr. K. Taira are acknowledged for their continuing encouragement, and Mr. Haiqing Li and Mr.Yihang Jiang for their support of the project. The project benefitted from constructive comments and suggestions by Dr. S. Jelgersma. We thank V.B.Bazarova, L.A.Ganzei, T.A.Grebennikova, A.M.Korotky, Haiquan Li, O.Ongkosongo, Y.Ono, R.Peape, V.S.Pushkar, N.G.Razjigaeva and Chuangli Xie for their active participation in the project. The Chinese National Natural Science Foundation and the Marine Geology Laboratory at the Tongji University supported the work in China and the final compilation of the maps. The authors are grateful to Prof. W.Berggren for improvement of the manuscript. Maiying Wu, Zhiwei Liu and Wei Huang are thanked for their technical assistance in preparation of the manuscript.

REFERENCES

1.   D.M.Anderson and R.S.Webb. Ice-age tropics revisited. *Nature*, 367, 23-24 (1994).
2.   Longxun Chen and Weiliang Li. The atmospheric heat budget in summer over Asia monsoon area. *Advances in Atmospheric Sciences*, 2, 487-497 (1985).
3.   Longxun Chen, ,Qiangen Zhu, Huibang Luo, Jinhai He, Min Dong, and Zhiqian Feng. *East Asian Monsoon*. China Meteorology Press, Beijing (in Chinese) (1991).
4.   K.Chinzei, K.Fujioka, H.Kitazato, I.Koizumi, T.Oba, M.Oda, H.Okada, T. Sakai, and Y. Tanimura, Postglacial environmental change of the Pacific Ocean off the coasts of central Japan. *Marine Micropaleontology*, 11, 273-291(1987).
5.   J.A Clark and C.S Lingle. Predicted relative sea-level changes (18000 years B.P. to present) caused by Last-Glacial retreat of the Antarctic ice sheet. *Quaternary Research*, 11, 279-298 (1979).
6.   CLIMATE Project Members. The surface of the ice-age earth. *Science*, 191, 1131-1137 (1976).
7.   B. Frenzel, M. Pecsi, and A.A.Velichko (eds.). *Atlas of Paleoclimates and Paleoenvironments of the Northern Hemisphere. Late Pleistocene - Holocene.* Geographical Research Institute, Budapest, and Gustav Fischer Verlag, Stuttgart (1992).
8.   M.G. Gross. *Oceanography: A View of the Earth* (4th edition). Pretice-Hall (1987).
9.   T.P Guilderson., R.G. Fairbanks and J.L.Rubenstone. Tropical temperature variations since the 20,000 years ago: modulating interhemispheric climate change. *Science*, 263, 663-665 (1994).
10.   Chi-Yu Huang, Sheu-Feng Wu, Meixun Zhao, Min-Te Chen, Chung-Ho Wang, Xia Tu and P.B. Yuan. Surface ocean and monsoon climate variability in the South China Sea since last glaciation. *Marine Micropaleontology* (in press).
11.   H.H.Lamb. *Climate: Present, Past and Future*. Methuan & Co., London (1972).
12.   Baohua Li, Zhimin Jian, and Pinxian Wang. *Pulleniatina obliquiloculata* as paleoceano-graphic indicator in the southern Okinawa Trough since the last 20,000 years. *Marine Micropaleontology* (in press).
13.   Kerang Li (ed.). *Climatology of the Seas near China and the North Pacific*. China Ocean Press, Beijing (in Chinese) (1993).
14.   Joo Tick Lim and Kwong Lun Tuen. Sea surface temperature variations in the South China Sea during the Northern Hemisphere winter monsoon. *Proceedings of the Second WESTPAC Symposium, 2-6 December 1991, Penang, Malaysia*, 113-144 (1991).
15.   B.K. Linsley, R.C.Thunell, C.Morgan, and D.F.Williams. Oxygen minimum expansion in the Sulu Sea, western equatorial Pacific, during the last glacial low stand of sea level. *Marine Micropaleontology*, 9, 395-418 (1985).
16.   J.I. Martinez. Late Pleistocene paleoceanography of the Tasman Sea: implications for the dynamics of the warm pool in the western Pacific. *Palaeo.,Palaeo.,Palaeo.*, 112, 1962 (1994).

17.   Q.Miao, and R.Thunell. Glacial-Holocene carbonate dissolution and sea surface temperatures in the South China Sea and Sulu Sea. *Paleoceanography*, 9, 269-290 (1994).

18.   T.C.Moore, Jr., L.H.Burckle, K.Geitzenauer, B.Luz, A.Molina-Cruz, J.H.Robertson, H. Sachs, C.Sancetta, J. Thiede, P.R Thomposon and C.Wenkam. The reconstruction of sea-surface temperatures in the Pacific Ocean of 18000 B.P. *Marine Micropaleontology*, 5, 215- 247 (1980).

19.   Yunqi Ni, and Yongfu Qian. The effects of sea surface temparature anomalies over the mid-latitude western Pacific on the Asian summer monsoon. *Acta Meteorologica Sinica*, 5, 28-39 (1991).

20.   T.Oba, M.Kato, H.Kitazato, I.Koizumi, A.Omura, T.Sakai, and T.Takayama. Paleoenvironmental changes in the Japan Sea during the last 85,000 years. *Paleoceanography*, 6, 499-518 (1991).

21.   Yunshan Qin, Fan Li, Baoyu Tang, and J.D.Milliman. Buried paleoriver system in western part of the South Huang Sea. *Kexue Tongbao*, 31: 1887-1889 (in Chinese) (1986).

22.   D.Rind, and D.Peteet. Terrestrial conditions at the last glacial maximum and CLIMAP sea-surface temperature estimations: are they consistent? *Quaternary Research*, 24: 1-22 (1985).

23.   I.Stuijts, J.C.Newsome and J.R.Flenley. Evidence for late Quaternary vegetational change in the Sumatran and Javan highlands. *Review of Paleobotany and Palynology*, 55: 207-216 (1988).

24.   J.Thiede. Wind regimes over the late Quaternary southwest Pacific Ocean. *Geology*, 7, 259-262 (1979).

25.   P.R. Thompson. Planktonic foraminifera in the western North Pacific during the past 150,000 years: comparison of modern and fossil assemblages. *Palaeo.,Palaeo.,Palaeo.*, 35, 241-279 (1981).

26.   P.R. Thompson and N.J. Shackleton. North Pacific palaeoceanography: late Quaternary coiling variations of planktonic foraminifer *Neogloboquadrina pachyderma*. *Nature*, 287, 829-833 (1980).

27.   R.Thunell, D.Anderson, D.Gellar, and Q.Miao. Sea-surface temperature estimates for the tropical western Pacific during the last glaciation and their implications for the Pacific Warm Pool. *Quaternary Research*, 41, 255-264 (1994).

28.   H.D.Tjia, S.Fujii, K.Kigoshi, A.Sugimuraand, and T.Zakaria. Radiocarbon dates of elevated shorelines, Indonesia and Malaysia, Part 1. *Quaternary Research*, 2: 487-495 (1972).

29.   T.Torgersen, M.F.Hutchinson, D.E.Searle, and H.A.Nix. General bathymetry of the Gulf of Carpentaria and the Quaternary physiography of Lake Carpentaria. *Palaeo.,Palaeo., Palaeo.*, 41, 207-225 (1983).

30.   UNESCO/IOC Western Pacific Subcommission. *WESTPAC Paleogeographic Maps*. The Last Glacial Maximum Paleogeographic Map for the Western Pacific Region. Shanghai (1995).

31.   J.H.F. Umgrove. *Structural History of the East Indies*. Cambridge. The University Press (1949).

32.   R.Vollbrecht and H.R.Kudrass. Geological results of a pre-site survey for ODP drill sites in the Sulu Basin. *ODP Proceedings, Initial Reports*, 124, 105-111 (1990).

33.   Luejing Wang and Pinxian Wang. An attempt at p[aleotemperature estimation in South China Sea using transfer function. *Chinese Science Bulletin*, 34, 53-56 (1989).

34.   Luejiang Wang and Pinxian Wang. Late Quaternary paleoceanography of the South China Sea: glacial-interglacial contrasts in an enclosed basin. *Paleoceanography*, 5, 77-90 (1990).

35.   Pinxian Wang, West Pacific marginal seas in last glacialtion: paleogeography and its environmental impact. *Proceedings of the Second WESTPAC Symposium, 2-6 December 1991, Penang, Malaysia*, 33-48 (1991).

36. Pinxian Wang, Response of West Pacific marginal seas to glacial cycles: paleoceanographic and sedimentological features (Abstract). *30th International Geological Congress, Abstracts*, **3**, 585.

37. Pinxian Wang and Rongfeng Li. Numerical modeling of sea surface circulation in the South China Sea at the last glaciation and its verification. *Chinese Science Bulletin*, **40**, 1813-1817 (1995).

38. Pinxian Wang and Xiangjun Sun. Last glacial maximum in China: comparison between land and sea. *Catena*, **23**, 341-353 (1994).

39. Pinxian Wang , Luejiang Wang, Yunhua Bian and Zhimin Jiang. Late Quaternary paleoceanography of the South China Sea: surface circulation and carbonate cycles. *Marine Geology*, **127**, 145-165.

40. P.J.Webster, and N.A.Streten. Late Quaternary ice age climates of tropical Australasia: interpretations and reconstructions. Quaternary Research, **10**, 279-309 (1978).

41. Jun Yan and P.R. Thompson. Paleogeographic evolution in the Okinawa Trough during the late Pleistocene. *Oceanologia et Limnologia Sinica*, **22**, 264-271 (in Chinese) (1991).

42. Xiaohai Yan, Chgungru Ho, Quanan Zheng and V.Klemas. Temperature and size variations of Western Pacific Warm Pool. *Science*, **258**: 1643-1745 (1992).

Proc. 30th Int'l. Geol. Congr., Vol. 13, pp. 87-94
Wang & Berggren (Eds)
©VSP 1997

Drilling mud volcanoes on the Mediterranean Ridge accretionary complex (ODP Leg 160)

ALASTAIR ROBERTSON and the Shipboard Scientific Party
Department of Geology and Geophysics, West Mains Road, Edinburgh University Edinburgh EH9 3JW, U.K.

Abstract

Two mud volcanoes, the Milano and Napoli mud volcanoes, were drilled during Leg 160 in the Eastern Mediterranean Sea south of Crete. These mud volcanoes are located towards the rear part of the Mediterranean Ridge, a mud dominated accretionary complex, created by subduction of Neogene to Holocene deep-sea sediments of the African plate beneath the Eurasian plate to the north. Only by drilling could the age and subsurface structure of the mud volcanoes be determined. The main results were that both mud volcanoes were periodically active for >1 m. yr. and that they are dominated by multiple debris flows composed of fragments of mainly sandstone and limestone in a muddy matrix. The most probable origin of the matrix is that it was at least partly derived from overpressured fluid-rich muds of late Miocene (Messinian) age located beneath the Mediterranean Ridge accretionary complex within the subduction decollement zone. By contrast, the lithified clasts were mainly derived from middle Miocene strata that probably by then formed part of the overlying accretionary complex. The recognition of a debris flow origin of "mud breccias" changes earlier views of these as viscous mud intrusions.

Keywords: Mediterranean, mud volcanoes, accretionary wedge, Ocean Drilling Program

Shipboard Party: A.H.F. Robertson, K.-C. Emeis (Co-Chief Scientists), C. Richter (Staff Scientist), M.-M. Blanc-Valleron, I. Bouloubassi, H.J. Brumsack, A. Cramp, G.J. Di Stefano, R. Flecker, E. Frankel, M.W. Howell, T.R. Janecek, M.-J. Jurado-Rodriguez, A.E.S. Kemp, I. Koizumi, A. Kopf, C.O. Major, Y. Mart, D.F.C. Pribnow, A. Rabaute, A.P. Roberts, J.H. Rullkotter, T. Sakamoto, S. Spezzaferri, T.S. Staerker, J.S. Stoner, B.M. Whiting and J.M. Woodside.

INTRODUCTION

At the time of the International Geological Congress in Beijing the P.R. China was seriously considering becoming a member of the Ocean Drilling Program. An I.G.C. symposium was devoted to ODP results and was enthusiastically supported especially by delegates from P.R. China. The first author gave a paper summarising the results of drilling two mud volcanoes on the Mediterranean Ridge south of Crete. This was one part of Leg 160, the first of two legs drilled in the Mediterranean Sea during the spring and summer of 1995 (Fig. 1). Leg 160 had both tectonic and paleoceanographic objectives. The paleoceanographic objectives concerned the origin of Pliocene to Holocene deep-sea organic-rich muds, known as sapropels. The tectonic objectives were twofold. The first concerned processes of collision of the Eratosthenes Seamount, a late Mesozoic to Neogene, carbonate platform with southern Cyprus along the active plate boundary of the African and Eurasian plates in the Eastern Mediterranean Sea. The second objective, the one outlined here, concerned the origin of unusual mud volcanoes located on the Mediterranean Ridge, a mud-dominated accretionary wedge created by subduction of the African plate beneath the Eurasian plate during the Neogene-Holocene. Ten days were spent on this objective. Following Leg 160, Leg 161 took place in the Western Mediterranean Sea and was concerned with completing the study of sapropels and with investigation of extensional crustal processes related to opening of the Alboran Sea.

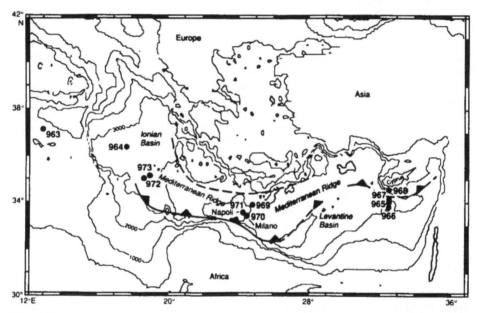

Fig. 1 Tectonic setting and location map of the Milano and Napoli mud volcanoes Sites 970 and 971 studied during ODP Leg 160. The leg had both tectonic and paleoceanographic objectives. The tectonic objectives were to study the setting of the Eratosthenes Seamount south of Cyprus and the origin of the mud volcanoes, as discussed here. The paleoceanographic objective concerned the origin of organic-rich deep-sediments, known as sapropels, that accumulated throughout the Mediterranean deep basin during the last 5 Ma.

The short article that follows in an updated version of one that was actually written at sea and conveys some the sense of excitement of discovery in the ocean through deep-sea drilling. The Ocean Drilling Program offers an unique opportunity for co-operation and new discoveries by the international deep ocean science community. More details can be found in the Initial Reports Volume for Leg 160 [1] and related preliminary results [2].

Regional geological setting

During the late 1980s, Italian researchers discovered the existence of domal structures on the Mediterranean Ridge, 150 km south of Crete [3,4]. The Mediterranean Ridge (500 km by 100 km wide) is interpreted as an accretionary prism, that developed during up to the last 25 m. yrs. as a result of northward subduction of the African plate beneath Eurasia. Seismic reflection studies have revealed an area of the seafloor that is characterised by numerous mound-like structures, which, when piston cored, were found to contain strange gaseous muds with "mousse-like textures" and so-called mud breccias, composed of clasts of hard and soft rocks in a soft sandy and clay-rich matrix. The mounds were first interpreted as partly diapiric structures that were intruded upwards onto the seafloor as relatively viscous masses [3,4,5,6]. During 1993, an international UNESCO-sponsored team returned aboard the Russian research vessel, Gelendzhik to the area [7,8]. Using an underwater camera they discovered that at least one of the structures, the Napoli Dome, is currently venting fluid and is surrounded by bacterial mats. Associated cold seep communities include shelled organisms. Deep-towed seismic equipment alseeeeeo revealed mud flows that radiate from central vents. The question thus arose: are the mud structures dome-like viscous intrusions, or cones of erupted clast-rich muddy sediments?

Results from the Milano mud volcano

The first of the structures to be drilled was the Milano Dome in the east (Fig. 2A). Seismic reflection data reveal it to have a sombrero-like shape, with a central cone surrounded by sloping flanks passing out into normal seafloor. The seismic reflectors beneath the flanks of the dome dip inward, forming a bowl-shaped depression beneath the volcano. How had such an unusual structure formed? The strategy was to drill a transect of shallow holes (<200 m) from the outer flanks to the crest of the two structures. After coring several meters of deep-sea calcareous ooze, we recovered core after core of a very lumpy sediment with scattered clasts of hard and softer rock within a stiff sandy clay (Fig. 3). Beneath this, we recovered several thin layers of layered deep-sea sediment, which are in turn underlain by repeated layers of clay-rich sand, silt and gravel, some showing evidence of deposition by turbidity currents. We were surprised when the shipboard palaeontologists determined an age: the deep-sea sediment beneath the lowest of the gravel-rich layers was 1.75 million years old. The Milano structure has been active, at least periodically, over an extraordinarily long time period. Further information came from borehole logging using geophysical tools. The formation microscanner revealed a layered structure suggestive of repeated extrusion of muddy debris flows and the presence of indurated clasts up to half a meter in size.

Our initial impression was that the mud-rich breccias represented debris flows, erupted from the vent of a mud volcano. How far away could these have extended from the eruptive centre? To find out we positioned our next hole further from the flanks of the structure. We then drilled though sediment typical of the surrounding sea floor (i.e. hemipelagic sediments, carbonate ooze and sapropels), which allowed us to show that mud eruptions had not flowed more than ca. 1 km from the eruptive centre of the Milano mud volcano. We then moved closer to the mud structure. Coring of the inner flanks revealed additional evidence of mud debris flows, whereas the crestal area was composed of more sandy material that might have formed a central plug-like structure.

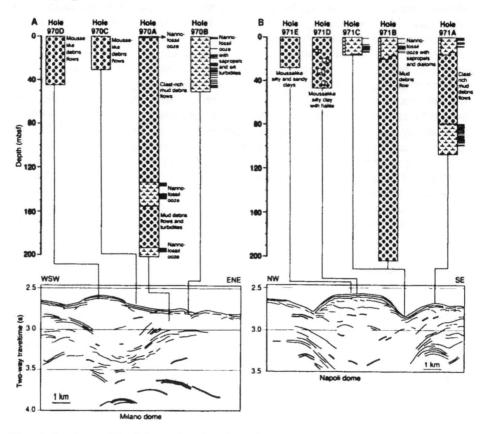

Fig. 2 Summary of the lithostratigraphy of (A) the Milano mud volcano and (B) the Napoli mud volcano drilled during Leg 160. The seismic reflectors visible within the two mud volcanoes are shown below each mud volcano. Note the presence of inward-dipping reflectors beneath both flanks of the Milano and Napoli mud structures, that are taken to indicate progressive subsidence during mud volcanism.

Further surprises were in store when the shipboard geochemists processed their data. Some of the cores from the crestal site were obviously gassy. An ice-like mixture of methane and water, known as clathrate (i.e. gas hydrate), must exist at depths of 30 or 40 m below the seafloor. The solutions from the crest were extremely low in salinity, indicating that clathrates had decomposed during the process of coring and bringing to the surface. In contrast, at deeper levels interstitial fluids were always much more saline than Mediterranean bottom water: this is because of the dissolution of Messinian evaporites that are assumed to exist beneath. Such salt was extensively precipitated during the late Miocene, about ~6-5 million years ago when the Mediterranean sea-level is known to have been at a much lower level than today [9].

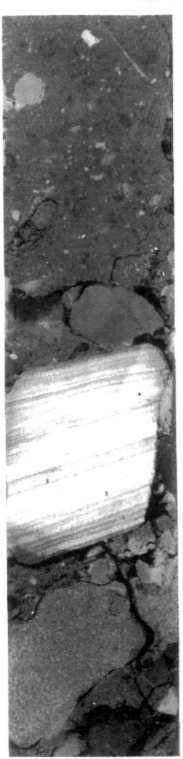

Fig. 3 Typical example a thick, relatively homogeneous mud debris flow from the the Milano mud volcano. The upper part shows the typical texture of the "mud breccias", interpreted as debris flows. The pale clast (centre) is quartzose sandstone, interpreted as a turbidite. The homogeneous clast near the bottom of the photograph is massive litharenite (sandstone). Site 970A, Core 10X, Section 1 at 56-82.5 cm.

Results from the Napoli mud volcano

We again adopting a strategy of drilling a transect of holes (Fig. 2B). We first positioned the drill string on the outer edge of the flank hoping to core and thus date the feather edge of the mud volcanic cone. We were right on target: beneath thin deep-sea sediments, we drilled through mud breccias, similar to those of the Milano mud volcano, then straight into normal deep-sea sediment, dated as 1.5 -1.2 Ma or older. To try to find out how thick these breccias are, we then drilled into a moat-like depression near the base of the central cone. After some 180 m with very little core recovery, we were still in mud-rich debris flows. At this point, we were alerted by a distinct whiff of hydrocarbons: a sample placed under ultraviolet light fluoresced strongly. We faced a dilemma: would we have to shut down the hole for safety reasons? However, there was no evidence of actively migrating oil and we were able to continue to our maximum planned depth. Hole conditions were, however, poor and our attempt to log this hole geophysically was only partially successful. However, we did find evidence of distinct layering in places within the mud breccias that we again interpreted as repeated mud flows.

We then proceeded to the summit of the crestal area. There the sediments were gassy and contained pungent hydrogen sulphide gas. Safety precautions were in place, although the volume of gas was high; its composition indicated a relatively shallow origin related to bacterial degradation. This did not constitute a safety hazard and we were able to core for an additional few tens of meters as planned. The muds that we recovered soon became frosted with crystalline salt as they started to dry, while some even contained small pieces of crystalline halite, of presumed Messinian age. Next, we repositioned the ship to core in the crestal area where active venting had been observed during the 1993 Gelendzhik expedition. Coring began normally, until the third core arrived on the deck: then before it could be properly processed, it ruptured explosively. Mud splattered all around! The cause was very rapid increase in the volume of gas due to rising temperature. Fortunately, nobody was injured, although showers and a change of clothing were in order. This was a real safety hazard. With reluctance, but knowing we had no choice, we "pulled pipe" and this exciting phase of our exploration of the Eastern Mediterranean mud volcanoes had come to an end. Luckily, however, this was the last of our planned drill holes and thus little information was lost.

DISCUSSION

We now know that the Milano dome is a submarine mud volcano that began to form at least 1.5 million years ago, while the Napoli structure is at least 1.5-1.2 Ma old. (Fig. 4 A, B). The inward dip of the seismic reflectors towards the volcanic centre suggests that progressive collapse of the volcano cone has taken place. Early eruption constructed a cone of unstable clastic sediment, including muddy debris flows and turbidites. Voluminous outpourings of mud flows then followed, interspersed with pelagic accumulation, eventually constructing the present cone.

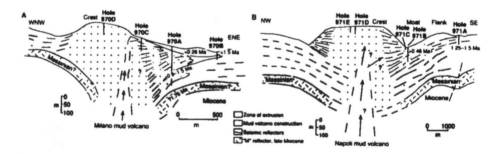

Fig. 4 Interpretation of (A) the Milano mud volcano and (B) the Napoli mud volcano. Muddy debris flows were erupted from central vents and accumulated within and adjacent to a subsiding moat.

Gas hydrates can form only in a limited temperature/pressure window and would not be stable if relatively warm fluids were escaping from depth. Consistent with this, gas hydrates are found associated with the inactive Milano mud volcano. By contrast, the Napoli volcano which is actively venting fluids and gases appears to lack gas hydrates.

What then was the driving force in mud volcanism near the northern edge of the Mediterranean Ridge? The deep Mediterranean basin in this region, a remnant of the Mesozoic Tethys ocean, is known to be in the last stages of closure and collision of

North Africa and Eurasia, such that the northern edge of the accretionary wedge is being thrust northward over continental crust south of Crete. The decollement zone is a possible location of overpressured fluid-rich sediment. Normally such material is prevented from moving upwards by a thick pile of previously accreted deep-sea sediments above. However, when backthrusting of the Mediterranean Ridge was initiated this created pathways for overpressured material to escape upwards onto the seafloor (Fig. 5). The probable source of a least part of the matrix was unfossiliferous mud of Late Miocene (Messinian) age located within the decollement zone. The probable origin of the clasts of limestone and sandstone is that they were detached from overlying Miocene accreted sediments, by a combination of hydraulic fracturing and physical abrasion. The mud- and and clast-rich material was then entrained upwards and eventually erupted on the seafloor as multiple debris flows. Post-cruise work is continuing to test and amplify this hypothesis.

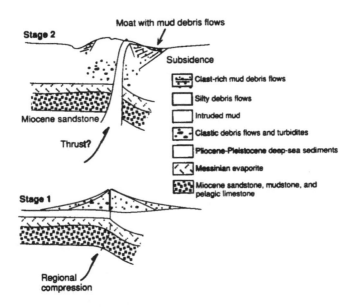

Fig. 5 Stages in the development of Mediterranean Ridge mud volcanism. (1) Early eruption build up a clastic cone and (2) Multiple debris flows build up the main structure of the mud volcanoes combined with progressive subsidence of a peripheral moat.

CONCLUSIONS

We gained important new insights on deep-sea mud volcanism from our ten day exploration of the Eastern Mediterranean mud domes. We learned that the Milano and Napoli mud domes are submarine mud volcanoes, perhaps similar to others in convergent margin settings, such as the Barbados subduction complex. As with magmatic volcanoes, these mud volcanoes appear to have a cyclic evolution that extends over several million years. Many more mud volcanoes probably remain to be discovered in other tectonically active areas of the ocean floor. Undoubtedly, other marine scientists will return before long to continue the exploration of this exotic geologic phenomenon.

Acknowledgements

As with all deep sea drilling legs nothing could have been achieved without the support of the captain, the crew and the marine technicians.

REFERENCES

1.   Emeis, A. H. F. Robertson, C. Richter and the Shipboard Scientific Party. *Initial Results of Ocean Drilling Program*, 160, College Station, Texas, Ocean Drilling Program (1996).
2.   Robertson, K.-C. Emeis (Co-Chief Scientists), C. Richter (Staff Scientist), M.- M. Blanc-Valleron, I. Bouloubassi, H.J. Brumsack, A. Cramp, G.J. Di Stefano, R. Flecker, E. Frankel, M.W. Howell, T.R. Janecek, M.-J. Jurado-Rodriguez, A.E.S. Kemp, I. Koizumi, A. Kopf, C.O. Major, Y. Mart, D.F.C. Pribnow, A. Rabaute, A.P. Roberts, J.H. Rullkotter, T. Sakamoto, S. Spezzaferri, T.S. Staerker, J.S. Stoner, B.M. Whiting, and J.M. Woodside, 1995, Mud volcanism on the Mediterranean Ridge: Initial results of Ocean Drilling Program Leg 160, Geology, 24, 239-242 (1996).
3.   Cita, A. Camerlenghi, E. Erba, F.W. McCoy, D. Castradori, A. Cazzani, G. Guasti, M. Giambastiani, R. Lucchi, V. Nolli, G. Pezzi, M. Redaelli, E. Rizzi, S. Torricelli and D. Violanti, Discovery of mud diapirism in the Mediterranean Ridge - A preliminary report, *Boll. Soc. Geol. It.* 108, 537–543 (1989).
4.   Camerlenghi, M.B. Cita, W. Hieke and T.S. Ricchiuto. Geological evidence of mud diapirism on the Mediterranean Ridge accretionary complex, *Earth Planet. Science Lett.* 109, 493–504 (1992) .
5.   Camerlenghi, M.B. Cita, B. Della Vedova, N. Fusi, Mirabile L. and G. Pellis, Geophysical evidence of mud diapirism on the Mediterranean Ridge accretionary complex, *Marine Geophys. Res.* 17, 115–141 (1995).
6.   Staffini, S. Spezzaferri and F. Aghib. Mud diapirs of the Mediterranean Ridge: sedimentological and micropalaeontological study of the mud breccia. *Riv. Ital. Paleont. Strat.* 99, 225–254 (1993).
7.   Limonov, J.M. Woodside and M.K. Ivanov (Eds). *Mud volcanism in the Mediterranean and Black Seas and shallow structure of the Eratosthenes Seamount.* Initial results of the geological and geophysical investigations during the third "Training through-Research" Cruise of the R/V 'Glendzhik', (June-July, 1993). UNESCO Reports in Marine Sciences, 64, 173 p. (1994).
8.   Premoli-Silva, E. Erba, S. Spezzaferri, and M.B. Cita, Age variation in the source of the diapiric mud breccias along and across the axis of the Mediterranean Ridge accretionary complex. *Marine Geology*, 132, 175-202 (1996).
9.   Hsü, M.B. Cita and W.B.F. Ryan, 1973, The origin of the Mediterranean evaporites. *In*: *Initial Reports of the Deep Sea Drilling Project*, 13, U.S. Government Printing Office, Washington, D.C., 1203–1231.

Proc. 30th Int'l. Geol. Congr., Vol. 13, pp. 95-110
Wang & Berggren (Eds)
©VSP 1997

Formation of Salt-marsh Cliffs in an Accretional Environment, Christchurch Harbour, Southern England

SHU GAO[1,2] and MICHAEL COLLINS[1]

[1] Department of Oceanography, The University, Waterfront Campus, Southampton SO14 3ZH, U.K.
[2] Institute of Oceanology, 7 Nanhai Road, Qingdao 266071, China

Abstract

Salt-marsh cliffs are widely distributed along the English coastlines. It has been proposed elsewhere that these features represent coastal erosion in response to periodic changes in hydrodynamic conditions. However, an example from Christchurch Harbour, southern England, shows that such cliffs can be developed within an accretional environment. Results derived from a mathematical model for simulating accretional processes and resultant morphological changes over an intertidal flat show that the cliff cannot be formed if, over the entire flat consisting of a mud flat and a sand flat, deposition rate at the mud/sand boundary is similar to the rate over the sand flat. However, if the difference between the two rates is sufficiently large, then the seabed slope at the sand-mud boundary will increase to such an extent that wave breaking takes place here. Further growth of the salt marsh, in combination with the breaking of waves, will lead to the formation of a cliff. The cliff may or may not retreat to landward, depending upon the intensity of the localised scour caused by breaking waves. In addition to such processes, sea-level rise may create the condition for the formation of salt-marsh cliffs in an environment where no cliffs will evolve under a constant sea-level; the presence of salt-marsh cliffs may represent a signal of sea-level rise. Using the magnitude of the deposition rate over the sand flat as a criterion, a salt-marsh cliff caused by sea-level rise may be identified.

Keywords: intertidal flats, accretional processes, wave action, suspended sediment transport, salt-march morphodynamics, sea-level rise

INTRODUCTION

Salt-marsh cliffs are widely distributed along the coastlines of England. They occur also in other parts of western Europe [6, 11] and also in eastern North America [13, 33], particularly within estuaries and coastal embayments. In order to explain the formation of the cliffs, a conceptual model characterised by accretion-erosion cycles has been proposed [2, 3]. According to this model, the cliffs are formed in response to accretion-erosion cycles controlled by periodic changes in external forcing (e.g. wave climate). When the condition favours accretion, salt marshes will advance to seaward. Likewise, when coastal erosion takes place in response to hydrodynamic changes, the salt marsh will retreat, producing a marginal cliff and an erosion surface over the lower intertidal flat. The accretion-erosion cycle may repeat itself to form a series of marshes with different ages. This model was based upon the study of the salt marshes in the Severn Estuary, southern England.

However, such a model may not provide a universal explanation for the formation of salt-marsh cliffs. It is assumed in the model that salt-marsh cliffs are associated with coastal erosion. Sometimes periodic accretion and erosion do occur in response to, for example, sub-tidal channel migration [30]). In other cases, the periodic changes in hydrodynamic conditions do not appear to exist. Further, although the cliff itself represents a scouring morphology, the scour may merely represent a localised phenomenon, rather than coastal erosion (i.e. a net loss of sedimentary material from the coastal system). In this paper, an example from Christchurch Harbour, southern England, shows that salt-marsh cliffs can be formed in an accretional environment.

If the cliff does not necessarily develop in an eroding environment, then a fundamental question will be how it should be formed. To answer this question, a mathematical model is established on the basis of the information on deposition rates and wave effects over the intertidal flat and salt marsh, to simulate the evolution of the morphology of the salt-marsh surface in an accretional environment (i.e. Christchurch Harbour) through a number of numerical experiments. Furthermore, using the results derived from the numerical experiments, the effect of sea-level rise on the salt-marsh cliff is examined and the possibility of identifying signals of sea-level rise from the cliff morphology is discussed.

SALT-MARSH CLIFFS IN CHRISTCHURCH HARBOUR

The Study Area

Christchurch Harbour (Fig. 1) is located on the coastline of southern England. The drainage basin of the harbour is approximately 3135 km^2 in area, which receives a freshwater discharge of around 30 m^3 s^{-1} on average. Outside the harbour, the longshore drift of sand and gravel eroded from the nearby sea cliffs has led to the development of two spits, which have semi-enclosed the harbour to form an estuarine / tidal inlet system [32]. The tidal basin has an area of 1.9 km^2 at MHWS (Mean High Water Spring), with a tidal prism of 1.43×10^6 m^3 under mean spring tidal conditions [16]. Within the harbour, the water is shallow, with a water depth of less than 2 m over the majority of the estuarine areas [25].

Salt marshes are distributed extensively over the bay-head areas (Fig. 1). The salt marshes are bordered by a low cliff, 0.4 to 0.7 m in height. The deposit exposed on the face of the salt-marsh cliff is characterised by fine-grained material; in the front of the cliff and over the lower intertidal flat, the bed material consists mainly of sands [15].

The harbour itself is well-sheltered from open sea waves. Over the offshore areas of Christchurch Bay, southwesterly waves are the prevailing and dominant waves. Here, the most frequently occurring wave heights are less than 0.6 m, but waves with a significant wave height of up to 7 m can occur [21]. Near the entrance to Christchurch Harbour, the largest wave heights have been observed to be reduced considerably, because of the presence of an ebb tidal delta. The entrance is relatively narrow (i.e. 47 m in width); this hinders the propagation of offshore waves into the harbour. The fetch for wind wave generation within the harbour is very small (i.e. < 1.2 km in any direction). Hence, the waves here are small; waves of 0.5 m in height can occur only during storms.

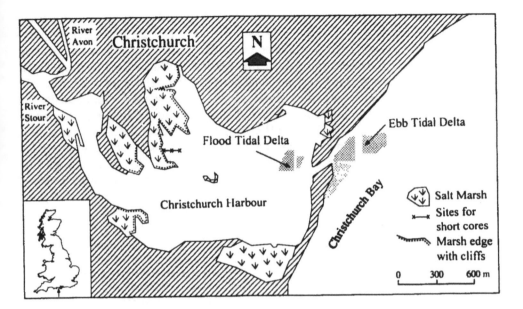

Figure 1. Distribution of salt marshes and salt-marsh cliffs in Christchurch Harbour (the location of the profile along which the short cores were collected are also shown).

The tides over the area are micro-tidal. At the entrance to Christchurch Harbour, mean spring and neap tidal ranges are only 1.4 m and 0.8 m, respectively. The mean flood and ebb tidal current speeds have been observed to be 0.17 m s^{-1} and 0.56 m s^{-1}, respectively [16]. Maximum cross-sectional mean tidal current speeds during the ebb can reach more than 2.0 m s^{-1}, when combined with a large freshwater discharge.

Field Survey and Sampling
Topographic survey along a typical intertidal flat profile and sediment sampling were undertaken within the harbour during 1989 to 1991 (for location of the profile, see Fig. 1).

The bed elevation along the intertidal flat profile was measured by levelling. The position of the western end and the orientation of the profile were determined using a DataScope digital compass. Along the profile, three short cores (0.56 m to 0.76 m in length) were collected. The sites of the short cores were 0 m, 50 m, and 100 m away from the western end of the profile, respectively. Further, 5 sediment samples were collected from the salt-marsh cliff face.

The marsh sediment samples and those taken from the short cores were analysed in laboratory, to determine the sediment type on the basis of the Folk [14] classification. The percentages of the mud, sand and gravel fractions were obtained using wet and dry sieving methods.

Sedimentological Characteristics

Analysis of the three short cores from the intertidal flat shows that the sediment sequence consists of a marine sand layer underlain by a set of gravelly deposits (Table 1 and Fig.2). The gravelly layer is considered to be a river terrace deposit, with its surface representing the base of Holocene deposits [5]. The marine deposit is characterised by white, light yellow and light grey sands, containing foraminiferal tests. The river terrace deposit consists of black and dark grey sandy gravel, rich in organic matter. Consequently, the boundary between the two layers represents the commence of the Holocene (marine or brackish) sediment deposition. Thus, the Holocene sediment cover over this area is less than 0.5 m over the lower part of the intertidal flat (Fig. 2). In terms of the samples from the cliff face, inorganic mud is dominant over the lower part of the cliff face, whilst the content of organic matter (represented by humus, grass roots and plant debris) increases towards the cliff top.

Table 1. Sedimentary sequence revealed by the short cores (C1, C2 and C3 are located at 0, 50, and 100 m from the salt-marsh cliff, respectively)

Core	Layer	Depth (m)	Gravel (%)	Sand (%)	Mud (%)	Sediment Characteristics
C1	1	0.00-0.07	0.0	86.6	13.4	Light yellow and grey sand
	2	0.07-0.41	0.0	96.2	3.8	White, clean sand
	3	0.41-0.50	5.4	88.6	6.0	White, clean sand and gravel
	4	0.50-0.76	0.0	94.3	5.7	Black sand
C2	1	0.00-0.16	6.1	79.4	14.5	Grey muddy sand
	2	0.16-0.29	0.3	93.7	6.0	Light yellow sand
	3	0.29-0.46	0.5	92.9	6.6	White and light yellow sand
	4	0.46-0.50	0.7	91.3	8.0	Black sand
	5	0.50-0.68	54.8	40.9	4.3	Black sandy gravel
C3	1	0.00-0.12	0.2	80.6	19.2	Muddy yellow and grey sand
	2	0.12-0.24	7.5	75.0	17.5	Grey gravelly muddy sand
	3	0.24-0.30	10.4	77.5	12.1	Yellow muddy sand and gravel
	4	0.30-0.47	54.6	43.0	2.4	Black sandy gravel
	5	0.47-0.56	53.4	43.0	3.6	Dark grey sandy gravel

The stratigraphic record does not appear to indicate the existence of an eroding surface, which would be predicted using Allen's model [2, 3] for salt-marsh cliff formation. A study using historical maps has shown that the shoreline within the harbour retreated only very slowly over the past decades [35]. Further, since the start of the present study in 1989, the position of the cliff edge along the profile has been monitored. During the past 7 years, the cliff retreated for less than 0.5 m. Hence, the retreat of the cliff edge at this particular location is a slow process; it does not involves any displacement of the whole profile.

Investigations into sediment dynamics of the inlet system have indicated that the harbour represents an accretional environment. The harbour receives sediment input from the

Rivers Avon and Stour. The discharge of fine-grained sediment is of the order of 8.5×10^6 kg yr^{-1}; this amount of suspended material is greater than that exported towards the sea through the harbour entrance [15]. In terms of the coarse-grained (sandy and gravelly) material, evidence from natural trace sediments [18, 36] and from grain size trends of seabed sediments [17] has shown that within the entrance long-term net transport is directed to landward. Therefore, the salt-marsh cliffs in Christchurch Harbour have been developed in an accretional environment.

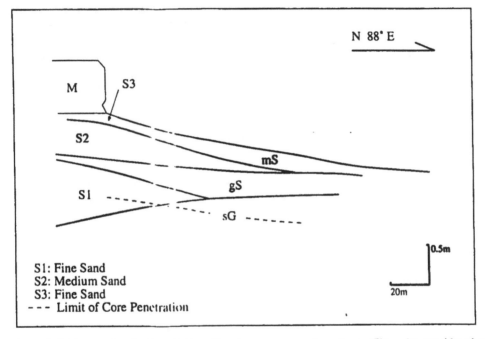

Figure 2. Holocene (marine/brackish) sedimentary sequence along the profile under consideration (the classification of sediments is based upon Folk [14]).

COMPUTER SIMULATION OF THE CLIFFS

Formulation of the Model
From the point of view of sediment dynamics, changes in the morphology of intertidal flats and salt marshes is controlled by the spatial gradient of net sediment transport rates. Hence, a possible method for the simulation of salt-marsh cliff development is to determine the spatial distribution pattern of the sediment transport rate. The application of such a method involves, however, the use of wave-induced and tidal current speed data which are not available for the study area. Thus, the approach adopted here, for a simulation study of the morphological evolution of a salt marsh, is to use directly the information on deposition rates, obtained in previous investigations and from the present study, in combination with an evaluation of wave effects on fine-grained sediment resuspension and movement over the intertidal zone.

An intertidal flat deposit may consist of two facies: an upper flat dominated by muds and a lower part of coarser material [12, 37]. In Christchurch Harbour, the stratigraphic record (see above) shows clearly such a "two-facies" structure. In order to simulate the morphological evolution, the distribution of deposition rates over these parts should be considered.

There have been extensive studies of the deposition rate over intertidal flats and salt marshes. The magnitude of accretional rates on a salt marsh may reach up to 25 mm yr^{-1}, but range generally between 0 and 10 mm yr^{-1} on temporal scales of less than 150 years, according to many investigations [e.g. 10, 24, 26]. The rate of erosion, in terms of changes in bed elevation, has been rarely reported (most studies (e.g. [9]) use the rate of horizontal retreat of the shoreline, in terms of m yr^{-1}). Generally, over the whole of the intertidal zone, a maximum deposition rate exists and the rate decreases towards both the land and the sea [24]. The maximum has been observed to be located near the lower limit of vegetation i.e. the boundary between the salt marsh and bare intertidal flat [33]. From this location towards the upper part of salt marshes, the deposition rate decreases [4, 28]. Over the sand flat, where sand supply is limited, the deposition rate over the lower part is significantly smaller than over the lower part of the salt marsh.

For a numerical expression of the salt marsh elevation growth, Pethick [28] has proposed that the bed elevation of a salt marsh can be fitted by an exponential model:

$$h = a - b\,e^{-ct} \tag{1}$$

where t is time, and a, b and c are constants. This implies that the deposition rate on the marsh surface (R_M) can be related to the bed elevation linearly:

$$R_M = \frac{dh}{dt} = \alpha + \beta h \tag{2}$$

where α and β are constants. According to Eq. 2, the deposition rate at the boundary of the sand flat and salt marsh (where h=0), R_0, is α. On the other hand, the physical meaning of the parameter β needs some explanation. Let T_0 be the length of time for a marsh to reach its mature stage and H_M be the associated elevation of the marsh surface. The following result can be derived from Eq. 2:

$$\int_0^{H_M} \frac{dh}{\alpha + \beta h} = \int_0^{T_0} dt \tag{3}$$

Thus, we have

$$\frac{1}{\beta} \ln\left(1 + \frac{H_0\,\beta}{R_0} \right) = T_0 \tag{4}$$

Eq. 4 means that β is a function of H_M, R_0 and T_0.

Further, the study by Pethick [28] showed that the deposition rate is high at the beginning of marsh developments (at the boundary of sand-flat/marsh facies), but it decreases rapidly within a period of 500 years. For Christchurch Harbour, H_M is taken here as the maximum marsh surface elevation (i.e. 0.7 m) and T_0 as 500 years (as suggested by Pethick [28]). In order to examine the sensitivity of α (i.e. R_0), different values should be used in the simulation experiments. Because Christchurch Harbour represents a micro-tidal environment, the deposition rate over the salt marsh should be relatively low (it has been observed elsewhere [20] that the rate increases with the tidal range). The stratigraphic record of the harbour shows also a low deposition rate over the Holocene period (Fig. 2). Therefore, the following three R_0 values are used in the computer simulation: 10, 7 and 3 mm yr^{-1}. Using any of these rates, a time-series of the bed elevation for the salt marsh with an increment of Δt can be defined on the basis of Eq. 2:

$$h_{k1} = h_i + (\alpha + \beta h_i) \Delta t \qquad (5)$$

Likewise, the change in the sand-flat elevation, over the time Δt, can be written as

$$h_{k1} = h_i + R_S \Delta t \qquad (6)$$

where R_S is the deposition rate over the sand flat. The information on the sedimentary history of the harbour (see above) reveals that the rate is lower than over the salt marshes. Thus, in order to examine the significance of R_S, different values will be used in the experiments (see below).

In addition to accretion over the marsh/flat surface, the wave conditions must be evaluated. It has been observed that changes in accretional/erosional patterns take place in response to seasonal changes in waves and wave-induced sediment resuspension. This is the case also for well-sheltered estuaries and coastal embayments where the fetch for wind-wave generation is small [1, 34].

Since no wave data exist for the harbour area, the evaluation of the wave effects here depends upon: (i) the calculation of wave climate, using relationships between wave parameters (height and period) and wave generating factors (fetch, wind speed and time); (ii) the determination of the location over the marsh surface where wave breaking occurs; and (iii) an assessment of the effect of wave breaking in terms of redistribution of fine-grained sediments and local scour, using information on suspended sediment concentrations and water depths.

A relationship between wave characteristics and the associated physical factors for shallow waters can be found in the *Shoreline Protection Manual* [8]. Using the maximum fetch of Christchurch Harbour (1.2 km), wave heights and periods under the various wind conditions are calculated. Using this information, together with the slope of the intertidal flat for the profile under consideration (i.e. 1/300), the water depths for wave breaking are also obtained (Table 2). The data show that the waves are small within the harbour, even under "stormy" meteorological conditions. The time required for the waves to reach an equilibrium state is relatively short, ranging between 9 and 30 minutes. Further, the calculated breaking depths mean that wave breaking in front of the marsh cliff takes place only during high water periods. These characteristics are consistent with field observations made on a number of occasions associated with strong winds.

On the basis of the study undertaken by Hydraulics Research [21], the probability of the occurrence of rough sea conditions (with a significant wave height of greater than 1.0 m over the offshore areas of Christchurch Bay, outside the harbour) is on average 31 % (or 113 days within a year). Further, large waves can be generated near the salt-marsh areas only by easterly winds, but the predominant winds for the region are from the southwest. Thus, the period when the study location is influenced by "stormy" events should be smaller than 113 days annually. Taking these factors into account, it is assumed that along the salt-marsh profile storm events occur during 75 days (or 150 tidal cycles) annually. For each of the tidal cycles, fine-grained sediment is assumed to be re-suspended and removed by a combined action of breaking waves and tidal currents.

Table 2. Wave characteristics calculated for Christchurch Harbour, under various wind speeds

Wind Speed (m s⁻¹)	Wave Height (m)	Wave Period (s)	Equilibrium Time (min)	Breaking Depth (m)
10	0.06	1.0	30	0.08
20	0.15	1.3	18	0.20
30	0.22	1.5	16	0.28
40	0.27	1.7	13	0.37
50	0.35	1.8	11	0.45
60	0.42	1.9	10	0.54
70	0.50	2.0	9	0.64

As shown in Table 2, except during the high water, waves generated under storm conditions break before they reach the cliff. Hence, the bed in front of the cliff will be affected by breaking waves, preventing the fine-grained material from accumulating (this is the case according to field observations). This means that the mud which will otherwise accumulate at this elevation at the rate of R_0 if the cliff is not present is removed due to the breaking waves. The volume of the fine-grained sediment removed annually and per unit width of the shoreline by the waves can be calculated by

$$V = \frac{1}{\lambda} N L C_s H_w \tag{7}$$

where λ is the bulk density of fine-grained sediment (taken as 1500 kg m⁻³), N is the number of tidal cycles associated with storm events, L is the range of the flat over which the removal of fine-grained material takes place (taken as the distance between the cliff edge and the boundary dividing the sand flat and salt marsh), C_s is the suspended sediment concentration, and H_w is the water depth (taken as 0.7 m, which is equivalent to the maximum height of a mature marsh cliff observed in Christchurch Harbour) within the range L. On the basis of field measurements, the suspended sediment concentration within the entrance to the harbour has a maximum of around 75 mg l^{-1} during storms [15]. At the study site, the concentration is anticipated to be higher. In the present study, therefore, C_S values of 0.1 and 0.2 kg m^{-3} are used.

The volume of the material available for the temporal erosion during storms, over L and per unit width of the shoreline, can be written as

$$V_F - \int_0^L h\, dx \tag{8}$$

where h is the bed elevation (with the origin being at the mud-sand boundary).

If V is smaller than V_F, then there will be a net accretion over the range L, and the cliff face will not be involved in erosion during storms. Hence, the cliff does not retreat in this case. On the other hand, If V is greater than V_F, then the cliff will retreat towards the land at a rate of

$$R_H - \frac{V - V_F}{H_C\, \Delta t} \tag{9}$$

where H_C is the height of the salt-marsh cliff in relation to the bed in front of the cliff.

Based upon the above equations, the position, height and the rate of cliff edge retreat can be modelled. For such an investigation, a Fortran program is written. The procedure of the simulation is as follows. (i) A file including the input data of the initial profile at t=0 (for its specification, see below), in the form of bed elevations at 1 m horizontal intervals, is prepared. (ii) Parameters R_0, R_S, and C_S are specified (other parameters including T_0, H_M, β and N are constants). (iii) The change in bed elevations, over a unit time (taken as 1 yr), is calculated using Eqs. (6) and (7). (iv) Based upon the bed slope calculated from the bed elevation data, whether or not a cliff is formed is examined (a slope of 1/100 is used as a criterion for cliff formation). (v) If the cliff is not formed, then repeat the above three steps, for next unit time. (vi) If the cliff is formed, then the difference in the bed elevation is taken as the cliff height, and the distance of cliff retreat is calculated on the basis of Eqs. (8) to (10). (vii) The steps (ii) to (vi) are repeated until the salt marsh reach the mature stage (over 500 years). Such a procedure provides the information on the profile morphology evolution, from which the height and location of the salt-marsh cliff can be derived.

The initial profile is assumed to be a strait line, with a slope of 1:300 (i.e. the slope of the present-day sand flat at Christchurch). The maximum and minimum elevations, at the ends of the profile, are 0.5 m and -0.5 m, respectively; this means that the origin of the vertical coordinate lies at the middle of the profile. The origin of the horizontal coordinate is located at the landward end of the profile.

Results of the Simulation Experiments
The results of the simulations, using the various combinations of the input parameters, are listed in Table 3. In the experiments, the parameters T_0 and H_M are taken as constants. Therefore, for the R_0 values of 10, 7 and 4 mm yr^{-1} and on the basis of Eq. 4, the related β values are -0.014275, -0.009935 and -0.00535, respectively. Altogether, 18 cases are

evaluated, using the various combinations of the R_0, R_S and C_S data. Base upon the modelled morphological evolution of the salt marsh, the following phenomena may be observed.

Table 3. Results of the simulation of the height (H_C) and location (represented by the distance from the origin, L_C) of salt-marsh cliffs $(T_0 = 500 \text{ yr}, H_M = 0.7 \text{ m}, N = 150)$

Case	R_0 (mm yr^{-1})	R_S (mm yr^{-1})	C_S (kg m^{-3})	H_C (m)	L_C (m) t = 50 yr	L_C (m) t = 500 yr
1	10	0.5	0.1	0.49	126	126
2	10	2	0.1	0.49	133	133
3	10	8	0.1	0	-	-
4	10	0.5	0.2	0.70	93	63
5	10	2	0.2	0.70	100	72
6	10	8	0.2	0	-	-
7	7	0.4	0.1	0.70	120	118
8	7	2	0.1	0.70	131	130
9	7	5	0.1	0	-	-
10	7	0.4	0.2	0.70	82	26
11	7	2	0.2	0.70	88	34
12	7	5	0.2	0	-	-
13	4	0.1	0.1	0.70	95	51
14	4	1	0.1	0.70	106	77
15	4	3	0.1	0	-	-
16	4	0.1	0.2	0.70	37	0
17	4	1	0.2	0.70	47	0
18	4	3	0.2	0	-	-

Firstly, if R_S is close to R_0, then the salt-marsh cliff cannot develop (see Cases 3, 6, 9, 12, 15, and 18). The reason for this, as far as the experiments are concerned, is that when the two deposition rates are similar no steep slopes will be formed and the conditions of wave breaking will remain the same as on the initial profile. Consequently, the growth of the sand flat matches that of the salt marsh. The remainder cases show that the cliff will be formed if R_S is considerably smaller than R_0. Because of the difference in the deposition rate, the slope at the boundary between the sand flat and lower part of the salt marsh becomes so steep after a period of marsh growth that waves start to break here. Further growth of the marsh, in combination of the wave-induced localised scour, results in the formation of the cliff.

Secondly, the cliff, after its formation, may or may not retreat during the salt-marsh evolution. In Cases 1, 2, 7, and 8, the position of the cliff remains stable or it retreats only very slowly. However, Cases 4, 5, 10, 11, 13 and 14 represent relatively rapid cliff edge recession. In Cases 16 and 17, the cliff recession is so rapid that the salt marsh is eroded completely when the marsh reaches its mature stage (thus, a new cycle of salt marsh evolution starts). These situations are controlled by a combination of the deposition rate R_0 and the effect of wave breaking, represented in the modelling by the magnitude of the suspended sediment concentration associated with wave breaking. For instance, in Case 2, which is characterised by a high R_0 value and a low C_s value, the position of the cliff is stable and its height is only 0.49 m (this implies the formation of a new salt-marsh in front of the old marsh, in spite of the effect of wave breaking) (Fig. 3). In Case 8, as shown on Figure 4, the position of the cliff is stable but the cliff height is smaller than in Case 2. This difference is caused by a smaller R_0 value used for the Case 8 experiment. For Case 11, the R_0 value used is the same as that for Case 8, but an increase in C_s results in cliff edge recession, at a rate of 0.11 m yr^{-1} on average over a 450-year period (Fig. 5).

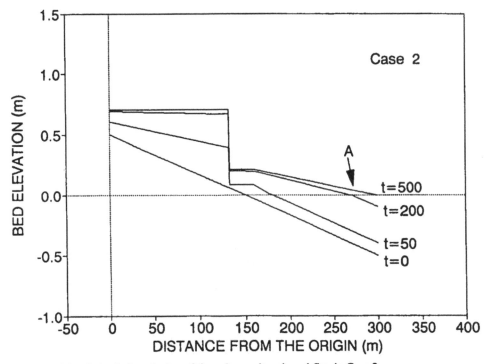

Figure 3. Morphological evolution of the salt marsh and sand flat, in Case 2.

Finally, as shown on Figures 3, 4, and 5, accretion occurs on the sand flat for the cases involved. Because the experiments are designed only for simulating the formation of salt-marsh cliffs, the effect of the sand flat growth is not modelled. However, it can be anticipated that at some stage the waves will no longer break over the bed in front of the cliff because of the widening of the sand flat. Instead, waves would break around the location "A" which is shown on the figures. In this case, another cycle for salt-marsh evolution starts: the sand flat will be transformed into marsh areas and a new cliff will develop in the vicinity of "A". It has been identified for a long time that sequential marshes with different ages and heights can develop within estuarine basins (e.g. [6]).

Hence, the simulation experiments show that the formation of salt-marsh cliffs represents an inherent characteristic in the evolution of a salt marsh in certain sedimentary environments. It does not imply necessarily coastal erosion, and the cliffs can be stable in terms of their position. Indeed, Carter [7] has pointed out that "marsh-edge cliffs may embrace aspects of both accretion and erosional activity, and should not be viewed automatically as an indication that the marsh is retreating".

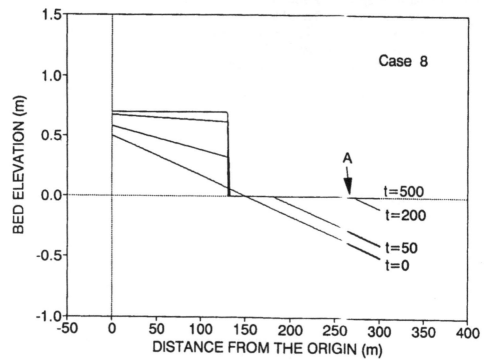

Figure 4. Morphological evolution of the salt marsh and sand flat, in Case 8.

DISCUSSION: SIGNAL OF SEA-LEVEL RISE

The present condition of sea-level rise has caused many concerns in terms of environmental changes (for a review, see [19]). In response to sea-level rise, salt marshes may be subjected to submergence, or they may remain stable and continue to grow, depending upon the amount of sediment supply [22, 23, 27, 29, 31]. Whatever the response is (i.e. survival or submergence), it is suggested here that the probability of the formation of a salt-marsh cliff will be enhanced due to sea-level rise.

On the basis of Eq. 2, the deposition rate is related linearly to the bed elevation, in terms of a fixed vertical coordinate system. However, sea-level rise is equivalent to a downwards movement of the coordinate system. In this situation, Eq. 2 should be rewritten as

$$R_M = \alpha \cdot \beta (h - R \Delta t) \tag{10}$$

where R is the rate of sea-level rise. Eq. 10 means that the accumulation of fine-grained sediment can be "rejuvenated" by sea-level rise (this hypothesis has been suggested also in [22]). Such a rejuvenation depends upon continuous supply of fine-grained sediment.

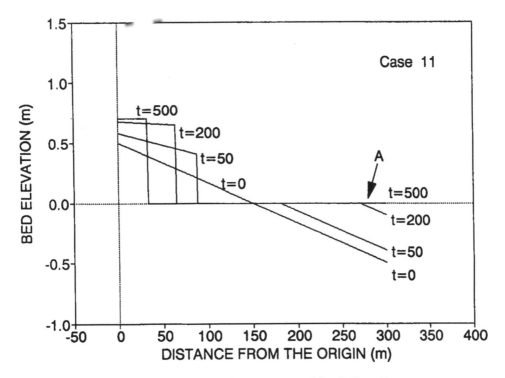

Figure 5. Morphological evolution of the salt marsh and sand flat, in Case 11.

On the other hand, in response to sea-level rise, the "apparent" deposition rate on the sand flat ($R_{s,a}$) will be modified, according to

$$R_{s,a} = R_s - R \tag{11}$$

As a result, $R_{s,a}$ will be smaller than the deposition rate without sea-level rise. It can be inferred, therefore, that sea-level rise enlarges the difference between the deposition rates of the lower salt marsh and the sand flat. This situation may, in turn, create the conditions for salt-marsh cliff formation according to the results obtained from the present study.

The observation that sea-level rise may cause the formation of salt-marsh cliffs implies that the presence of the cliff may contain some signals of sea-level changes. However, a question arises with regard to the differentiation of the cliffs formed under a constant sea-level and those introduced by sea-level rise. On the basis of the computer experiments presented above in which a constant sea-level is assumed, it can be inferred that there exists a critical value of R_s for any fixed R_0 value: if $R_s > R_{s,C}$ (where $R_{s,C}$ is the critical value), then the cliff will not be formed. However, when sea-level rise takes place, the cliff may be formed even if $R_s > R_{s,C}$, according to Eqs. 10 and 11. Hence, the parameter $R_{s,C}$ can be used as a criterion to distinguish the cliffs due to sea-level rise from those associated with a constant sea-level. The formation of any cliff where the deposition rate on the sand flat exceeds the critical value is indicative of the effect of sea-level rise. More investigations are required to determine the critical deposition rate.

CONCLUSIONS

An example from Christchurch Harbour, southern England, shows that salt-marsh cliffs can be formed in an accretional sedimentary environment. Hence, the presence of salt-marsh cliffs is not necessarily indicative of coastal erosion; it may only represent a phenomenon of localised scour.

For an accretional environment, a mathematical model is used to simulate the evolution of the salt-marsh morphology. The results show that the cliff will not be formed if maximum deposition rates over the marsh surface are similar to those over the sand flat. However, if the difference between the two is sufficiently large, then the seabed slope at the sand-mud boundary will increase to such an extent that wave breaking takes place here. Further growth of the salt marsh, in combination with wave breaking, will then cause a cliff to develop. The cliff may or may not retreat, depending upon the intensity of the localised scour caused by the breaking waves. In well-sheltered estuaries or embayments, the location of the salt-marsh cliff may be stable or retreat only slowly.

Sea-level rise may create the condition for the formation of salt-marsh cliffs in an accretional environment in which no cliffs will be formed if sea-level remains stable. Hence, the presence of such a cliff is indicative of sea-level rise. In order to identify such a category of salt-marsh cliffs, the magnitude of the deposition rate over the sand flat can be used as a criterion.

Acknowledgements

The authors wish to thank Adonis Velegrakis and Xiankun Ke for their assistance with field work, including sediment sampling and topographic survey. Mrs Kate Davis is thanked for her assistance with the preparation of some of the figures. During the study, SG was supported financially by the British Council and the Chinese Government, through a grant from the TC (UK-China Technical Cooperation) scheme, and the Chinese Academy of Sciences, through the "Hundred-Talent Program" (Marine Sediment Dynamics). Finally, Dr H.-S. Zhang (Institute of Oceanology, Chinese Academy of Sciences) is thanked for her assistance with the typing of the manuscript.

REFERENCES

1. F.E. Anderson. Resuspension of estuarine sediments by small amplitude waves. *Journal of Sedimentary Petrology*, **42**, 602-607 (1972).
2. J.R.L. Allen. Evolution of salt-marsh cliffs in muddy and sandy systems: a qualitative comparison of British west-coast estuaries. *Earth Surface Processes and Landforms*, **14**, 85-92 (1989).
3. J. R. L. Allen. The Severn Estuary in southwest Britain: its retreat under marine transgression, and fine-sediment regime. *Sedimentary Geology*, **96**, 13-28 (1990).
4. J.R.L. Allen. Shoreline movement and vertical textural patterns in salt marsh deposits: implications of a simple model for flow and sedimentation over tidal marshes. *Proceedings of the Geologists' Association*, **107**, 15-23 (1996).

5. C.R. Bristow, E.C. Freshney and I.E. Penn. *Geology of the Country Around Bournemouth*. Her Majesty's Stationery Office, London (1991).
6. A.E. Carey and F.W. Oliver. *Tidal Lands*. Blackie, London (1918).
7. R.W. Carter. *Coastal Environments: An Introduction to Physical, Ecological and Cultural Systems of Coastlines*. Academic Press, London (1988).
8. Coastal Engineering Research Center. *Shore Protection Manual (4th edition)*. US Government Printing Office, Washington, D.C. (1984).
9. R.A.Dalrymple, R.B. Biggs, R.G. Dean and H. Wang. Bluff recession rates in Chesapeake Bay. *Journal of Waterway, Port, Coastal and Ocean Engineering (ASCE)*, 112, 164-169 (1986).
10. O. Denema and R.D. DeLaune. Accretion rates in salt marshes in the eastern Scheldt, south west Netherlands. *Estuarine, Coastal and Shelf Science*, 26, 379-394 (1988).
11. M.M. van Eerdt. Salt marsh cliff stability in the Oosterschelde. *Earth Surface Processes and Landforms*, 10, 95-106 (1985).
12. G. Evans. Intertidal flat sediments and their environments of deposition in the Wash. *Quarterly Journal of the Geological Society of London*, 121, 209-245 (1965).
13. K. Finkelstein and C.S. Hardaway. Late Holocene sedimentation and erosion of estuarine fringing marshes, York River, Virginia. *Journal of Coastal Research*, 4, 447-456 (1988).
14. R.L. Folk. *Petrology of Sedimentary Rocks*. Hemphill Publishing Company, Austin (Texas) (1980).
15. S. Gao. *Sediment Dynamics and Stability of Tidal Inlets* [Unpublished Ph.D. Thesis]. University of Southampton, Southampton (1993).
16. S. Gao and M. Collins. Tidal inlet stability in response to hydrodynamic and sediment dynamic conditions. *Coastal Engineering*, 23, 61-80 (1994).
17. S. Gao and M. Collins. Analysis of grain size trends, for defining sediment transport pathways in marine environments. *Journal of Coastal Research*, 10, 70-78 (1994).
18. S. Gao and M. Collins. Net transport direction of sands in a tidal inlet, using foraminiferal tests as natural tracers. *Estuarine, Coastal and Shelf Science*, 40, 681-697 (1995).
19. V. Gornitz. Sea-level rise: a review of recent past and near-future trends. *Earth Surface Processes and Landforms*, 20, 7-20 (1995).
20. E.Z. Harrison and A.L. Bloom. Sedimentation rates on tidal salt marshes in Connecticut. *Journal of Sedimentary Petrology*, 47, 1484-1490 (1977).
21. Hydraulics Research. *Wind-Wave Data Collection and Analysis for Milford-on-Sea*. Hydraulics Research Limited Report No.EX1979, Wallingford (1989).
22. S.C. Jennings, R.W.G. Carter and J.D. Orford. Implications for sea-level research of salt-marsh and mudflats accretion processes along paraglacial barrier coasts. *Marine Geology*, 124, 129-136 (1995).
23. J.T. Kelley, W.R. Gehrels and D.F. Belknap. Late Holocene relative sea-level rise and the geological development of tidal marshes at Wells, Maine, USA. *Journal of Coastal Research*, 11, 136-153 (1995).
24. W.S. Letzsch and R.W. Frey. Deposition and erosion in a Holocene salt marsh, Sapelo Island, Georgia. *Journal of Sedimentary Petrology*, 50, 529-542 (1980).
25. J.W. Murray. A study of the seasonal changes of the water mass of Christchurch Harbour, England. *Journal of Marine Biological Association of the United Kingdom*, 46, 561-578 (1966).
26. M.M. Nichols. Sediment accumulation rates and relative sea-level rise in lagoons. *Marine Geology*, 88, 201-219 (1989).
27. R. Orson, W. Panageotou and S.P. Leatherman. Response of tidal salt marshes of the US Atlantic and Gulf coasts to rising sea levels. *Journal of Coastal Research*, 1, 29-37 (1985).
28. J.S. Pethick. Long-term accretion rates on tidal saltmarshes. *Journal of Sedimentary Petrology*, 51, 571-577 (1981).
29. J.D. Phillips. Coastal submergence and marsh fringe erosion. *Journal of Coastal Research*, 2, 427-

436 (1986).

30. A.W. Pringle. Erosion of cyclic saltmarsh in Morecambe Bay, north-west England. *Earth Surface Processes and Landforms*, **20**, 387-405 (1995).

31. D.J. Reed. The response of coastal salt marshes to sea level rise: survival or submergence. *Earth Surface Processes and Landforms*, **20**, 39-48 (1995).

32. A.W.H. Robinson. The harbour entrances of Poole, Christchurch and Pagham. *Geographical Journal*, **121**, 33-50 (1955).

33. J.-B. Serodes and J.-P. Troude. Sedimentation cycles of a freshwater tidal flat in the St. Lawrence Estuary. *Estuaries*, **7**, 119-127 (1984).

34. G.L. Shideler. Suspended sediment responses in a wind-dominated estuary. *Journal of Sedimentary Petrology*, **54**, 731-745 (1984).

35. P. Tosswell. *An Hydraulic Investigation of Christchurch Harbour with Particular Reference to the Entrance Channel* [Unpublished B.Sc. Dissertation]. Southampton University, Southampton (1978).

36. P.-X. Wang and J.W. Murray. The use of foraminifera as indicators of tidal effects in estuarine deposits. *Marine Geology*, **51**, 239-250 (1983).

37. Y. Wang and X.-H. Gu. The shell coastal ridges and the old coastlines of the west coast of the Bohai Bay (in Chinese, with English abstract). *Journal of Nanjing University (Natural Science Version)*, **8**, 424-442 (1964).

Proc. 30*th* Int'l. Geol. Congr., Vol. 13, pp. 111-128
Wang & Berggren (Eds)
©VSP 1997

Stratification of ferromanganese crusts on the Magellan seamounts

Irina Pulyaeva

Joint-Stock Company Dalmorgeologia, Nachodka, 375528, Russia.

Abstract

The conditions of bedding, structure, composition and age of ferromanganese crust of the Magellan seamounts have been examined. The generalised section of the Magellan seamounts crusts consists of three units. The lower unit I is characterised by the high degree of transformation of the substance and saturation with phosphate components. The lower zone I-1 and the upper zone I-2 of this unit are singled out according to textural characteristics. The upperlying units II and III are affected by diagenesis to a lesser degree, being characterised by increased concentration of useful components and decreased concentration of phosphates. Within the Magellan seamounts region the proto-unit R of ancient crusts was discovered in one per cent of geological stations. This relict of previous crusts differs from modern ones by structure, mineral and chemical composition. On the basis of biostratigraphy investigations it has become possible to distinguish six age groups of calcareous nannofossils recovered from the main elements of crusts section. Judging from the conditions of bedding, structure, composition and age of crusts on guyots of the Magellan seamounts, three generations of ferromanganese crusts can be recognised, corresponding to three stages of forming: late Cretaceous (unit R), Eocene (unit I) and Neogene - Pleistocene (units II and III). The section of ferromanganese crusts is broken off by numerous interruptions. The history of ferromanganese crusts formation is correlated with the evolution of transport of biogenic calcium carbonate to the ocean floor and its subsequent dissolution on the bottom. The main periods of ore substance accumulation coincide with the increase of the ocean productivity and evolutionary radiation of carbonate microplankton. In the periods of low productivity of the ocean regional interruptions in carbonate accumulation occur, they correlate with interruptions in crusts forming.

Keywords: guyots, Magellan seamounts, ferromanganese crusts, structure, generation, stratification.

INTRODUCTION

Nowadays oceanic ferromanganese crusts are of great interest as potential deposits of polymetallic ores. Much research on the structure and composition of the crusts of seamounts Wake-Nekker, the Hawaii archipelago, the Tuamotu archipelago, the Marshall Islands, guyots of IOAN and Ita-Maitaiya, Krylov Mount and some other objects have been done [2, 6, 7, 8, 9, 11, 12].

It has been determined that ferromanganese crusts, making up multy-unit ore formations, are the analogue of the sedimentary thickness and are of sedimentary origin. According to P.Halbach and D.Puteanus, skeleton remnants of carbonate plankton organisms are the main suppliers of iron to the surface of the ocean floor[7]. The school of marine geologists headed by, A.P. Lisitsin, Corresponding Member of the Academy of Sciences of Russia has developed main conceptions of the mechanism of delivery of iron to the sea bed surface in composition of planktonogenic calcium carbonate and its disengagement during its dissolution in the layer of oxygen minimum[2].

Newly precipitated oxide ferrominerals catalystically oxidize on their surface the manganese protoxide dissolved in the near - bottom water, and together they saturate the ore substance of the forming ferromanganese crusts with cobalt, nickel and copper as a result of their transition into solid phase from the surrounding oceanic water. Precipitation of metalcontaining suspended particles is possible on subhorizontal surfaces of solid substratum at the depth of 500 - 2500 m, i. e. in the water layer of oxygen minimum in the presence of subtle balance between the input of precipitation material on the bottom surface and the output of its considerable part by bottom streams from the area of its ore accumulation [2].

In order to reconstruct the history and determine regularities of ferromanganese orogenesis it is necessary to explore the problem of stratification of crust section. The problem of age of ferromanganese formations has been solved mainly by methods of absolute geochronology with episodic usage of paleomagnetic and biostratigraphic research [1, 5, 11]. The use of paleomagnetic method is subject to suspicion, because the character of breaks between the units of ferromanganese crusts is still not clear. The suggestion that the crusts up to now remain an open system casts doubt on the basis of the result of radioisotop methods of age determination and growth speed of ferromanganese crusts. Therefore, methods of biostratigraphy become especially significant, and the results of microfossil associations "in-situ" provide age definition of ferromanganese substance containing these associations.

The author of the present article carried out some biostratigraphical studies of the sedimentary rocks and ferromanganese crusts of the Magellan seamounts. On the basis of the received data we examined stratification of the main elements of crust section and correlated the history of ore formation with the history of geological development of seamounts as well as the global changes of paleooceanic characteristic. Materials of the research carried out by vessels of the Joint-Stock Company Dalmorgeologia on guyots IOAN, MG-36 and Dalmorgeologia have been used (Figure 1). The author took part in the work of these vessels, investigating geological structure of the guyots and outlined ore basins of ferromanganese crusts. Information on geological structure of guyots and conditions of bedding of ferromanganese crusts has been obtained from the data of multy-ray echosounding, seismoacoustic profiling, hydroacoustic, sonar and photo-TV surveys. Bottom sampling was completed using a 1-m-diameter circular chainbag dredge. Location of the vessel was determined with the help of satellite systems of navigation "GLONASS" and "NOVSTAR". Location of submarine apparatus with regard to the vessel was determined with the help of the short basis hydroacoustic navigation system "Rakoushka".

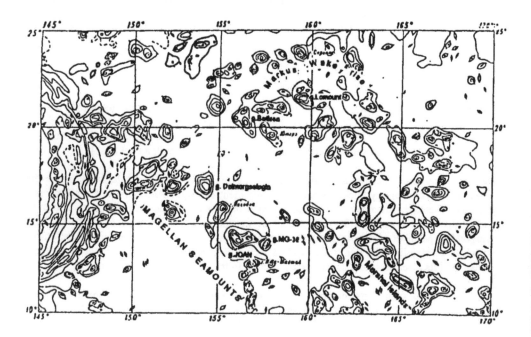

Figure 1. Scheme of the bottom relief at the Magellan Seamounts.

Chemical composition of crusts was studied in laboratories of the Joint-Stock Company Dalmorgeologia, the Institute of Chemistry of Far East Branch of the Russian Academy of Sciences and the Central laboratory of the Production Association Kamchatgeologia. Manganese was measured by potentiometric titration analysis; Fe, Co - by atomic absorption; P_2O_5 - by photocolorimetry (by A.I. Vasiliev, S.M. Balabon); Ni, Cu, Zn, Pb, As - by X-ray fluorescence (by V.M. Petrukhin); Pt, Au, Ag - by fire assay with AES analysis (by G.V. Borzina); REE, Mo - by X-ray radiometry analysis (by A.A. Terski).

Mineralogy composition was studied with X-ray diffractometer DRON-3,0 (by Ju.E. Makagonova, M.A. Budilko and M.E. Melnikov). Identification of microfossil assemblages was carried out by I.A. Pulyaeva with a scaning electronic microscope of JSM series.

GEOLOGIC SETTING

Ferromanganese crusts cover 90 per cent of the areas of outcrops of fundamental rock on the slopes and top parts of guyots. Volcanogenic and sedimentary formations of Cainozoic and Mesozoic take part in geological structure of guyots. Volcanogenic formations include two complexes. The first, the early one, composes pedestals of paleovolcanoes between isobaths 3000 - 5000 m. It is represented by lavas of toleitous basalts. The second complex constituting the upper part of guyots (to isobath 3000 m.) is marked by the variety of basalts and their increased alkalinity. Formation of volcanoes and their evolution from shield volcanoes to stratovolcanoes of the central type took place in the early Cretaceous (Barremian - Aptian).

The composition of sedimentary rock complex has proven the fact that by the middle Albian volcanoes, having subjected to the abrasion of upper parts, entered the atoll stage development which was interrupted by the Late Cretaceous transgression and submergence of guyots. The complex consists of:

- volcanogenic sedimentary rocks of Aptian - Albian - products of volcano destruction;
- reefal limestones of Albian - products of reef formation and reef destruction;
- planktonogenic carbonate sediments of the Late Cretaceous - Pleistocene - products of bathyal - pelagic conditions of sediment accumulation set on the slopes of guyots since the time of their submergence.

Of special interest are planktonogenic carbonate sediments, with coccollites and foraminifer skeleton remnants put together which are the main suppliers of iron to the surface of the ocean floor [2, 7].

The beginning of the guyot submergence dates to the Cenomanian. Sediments of this age are represented by tuffites with layers of nannoforaminiferal limestones and calcareous sandstones, formations of which are connected with existence of the vast surfy zones which were on the top parts of the submerging guyots. On the slopes of guyots there are also nannoforaminiferal limestones of Cenomanian - Turonian, formation of which took place at the depths of shelf - upper bathyal, which is proved by presence of complex of benthic and plankton foraminifer of the corresponding age discovered in them. These sediments fixed the beginning of the ruin of biocenoses of reef sediments and their erosion at the boundary of the Early and Late Cretaceous.

Sediments of Campanian - Maastrichtian age are widely distributed on the guyot slopes, and testify to the further submergence of guyots to the depths lower than the level of the ocean wave influence. They are represented by breccias. The cement of the latter is nannoforaminiferal limestones forming covers of their own on the levelled socle surfaces. The complex of plankton foraminifer discovered in them testifies to the bathyal conditions of sediment accumulation on the slopes of guyots in Maastrichtian. Fragments of ore substance of ferromanganese compound are present in the cement of the breccias.

Stabilisation of guyots at the present - day level took place in Eocene, and the data of deep sea drilling are evidence of it. According to these data, the lower layers of nannofossils ooze lying on oolite limestones of reefal complex date to the middle Eocene (site 202). Nannoforaminiferal limestones of Eocene form covers on top parts and slopes of guyots. They are intensively phosphotized and impregnated with manganese hydroxides. Breccias, the cement of which is limestones of Eocene, also contain numerous fragments of ferromanganese crusts.

During the Neogene - Pleistocene, formation of friable ooze takes place, and its thickness on the top parts is as much as 100-150 m., according to the data of seismoacoustic profiling. Inside the thickness, according to the results of hydroacoustic survey, characteristic lamination is exhibited on separate parts; it may be accounted for by the accumulation of thin interlayers of ore material.

It is not excluded that since the moment of the submergence of guyots to the bathyal depth and accumulation of nannofossil ooze on their slopes, favourable conditions are created for the formation of ferromanganese crusts.

The most advantageous conditions for the development of crusts are the sediment-free parts of guyots top plats. The example of this is the ore deposit with an area of about 100 km^2 outlined within the limits of the saddle of guyot IOAN. Ferromanganese crusts there form the cover stable enough from the point of view of its thickness and distribution. They lie on   volcanogenic sedimentary rocks (tuffites) and   limestones of the reefal complex. The average thickness of crusts is as much as 10 cm, according to the results of dredge testing confirmed by core drilling.

The most advantageous conditions for crusts development on the slopes are side spurs where surfaces of lava streams of basalts are almost even. On the steep parts of the slopes the thickness of ore covers decreases as well as their completeness is broken. On vertical ledges a thin film of ferrohydroxides and manganese hydroxides is found.

STRUCTURE OF FERROMANGANESE CRUSTS

It has been established that the thickness of crusts is the function of their structure the main feature of which is bedding. It is caused by parallel - streaky changes of structure - textural features in section [12]. The generalised section of the Magellan seamounts crusts shows three units (Figure 2).

Unit I beds directly on the substratum. Its thickness varies between 2 and 8 cm. The ore substance is bluish - black with metallic luster, shell - like fracture, firm, dense. Its density is 2,17 g/cm^3 , humidity is 22 per cent. In the structure of the unit two zones with a number of distinguishing structure - textural features are singled out. The ore substance of the lower zone (I - 1) is characterised by the thin - layer texture and the presence of conformable veins and cross veins composed with phosphatized nannoforaminiferal mass. The upper zone (I - 2) consists of ore substance with indistinct spotted columnar texture which is conditioned by the combination of ferrohydroxide and manganese hydroxide columns and phosphatized nannoforaminiferal mass filling the interstices between structural elements of ore substance. Zones I-1 and I-2 can be present in the section simultaneously. The contact between them is gradual.

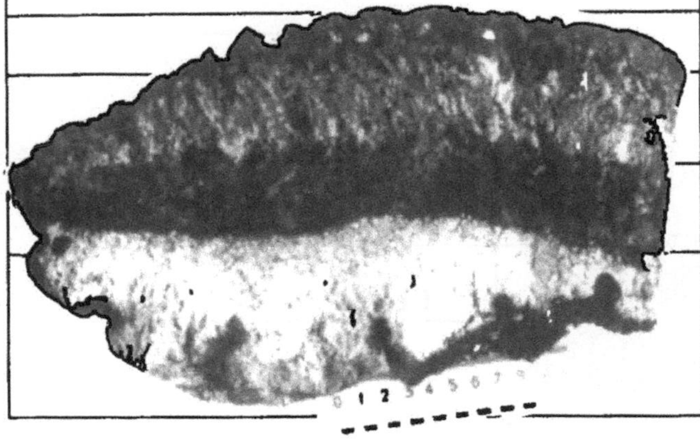

35 D 100 (D-1733 m)	Units	Age
	III	N_2-Q
	II	P_2^3-N_1
	I-1	P_1^2-P_2^1
		K_1a-al

Figure 2. Three unit section of crusts on reef limestones.

Higher in the section lies **unit II** with characteristic radial - columnar structure. It is 2 - 6 cm thick. It has constant regular contact with the substance of zone I-2. If there is no zone I-2 unit II lies on the substance of zone I-1 with a rather sharp contact sometimes with angular unconformity (Figure 3).

35 D 63 (D-1406 m)	Units	Age
	III	N_2-Q
	II	P_2^3-N_1
	I-1	P_1^2-P_2^1
		K_2

Figure 3. Three unit sections of crusts on volcanogenic sedimentary rocks.

The ore substance features high porosity, humidity is as much as 39 per cent, density amounts to as much as 1,78 g/cm^3. Ferrohydroxides and manganese hydroxides form vertically oriented columns of branchy dendrite shape. Interstices between them are filled with clayey, more rarely carbonate-clayey suspension.

Unit III, 1-4 cm thick, completes the section. The contact with unit II is gradual. The massive ore substance is black. The texture can be columnar, porous, spotted. Humidity is 32 per cent, density is 1,90 g/cm^3.

Approximately in 1 per cent of samples the main section is laid with relics of proto-unit R (Figure 4).

35 D 131 (D-2019 m)	Units	Age
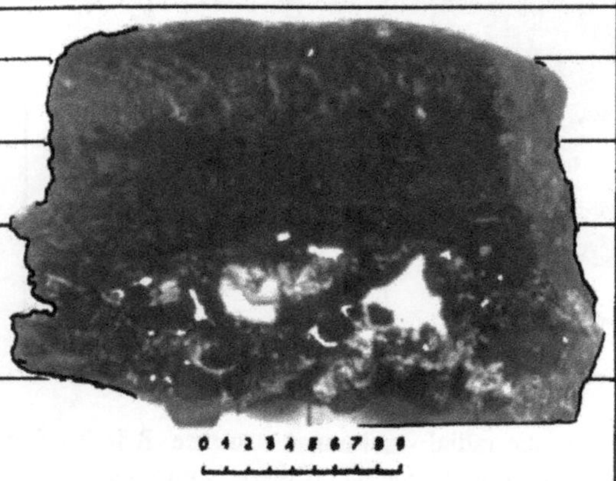	III	N_2-Q
	II	P_2^3-N_1
	I-1	P_1^2-P_2^1
	R	K_2
		K_1a-al

Figure 4. Three unit section of crusts on reef limestones.

In general terms, it is characterised by high density (2,46 g/cm^3) and low humidity (11 per cent), as well as by block structure of ore substance and high non-ore components saturation. The latter are represented by carbonate-phosphate material present in the form of syngenetical lenses, conformable veins and cross veins. The contact with the above mentioned units is sharp, underlined by streaks of phosphate material and basalt fragments. Carbonate-phosphate cement breccias of the Late Cretaceous sometimes contain ore fragments analogous to the ore substance of unit R.

On the whole the generalised crust section is divided into units with high degree of diagenetical transformation of ore substance (R and I) and units (II and III) affected by the process of diagenesis in a considerably lesser degree.

The occurrence of the complete section is different on each guyot, e.g. on guyot IOAN it comes to 41 per cent, and on guyot Dalmorgeologia it is 30 per cent. In other cases crusts have the reduced section represented by units II and III, or only by unit III; more rarely occur combinations of units I and III.

It has been determined that the maximum frequency of three-unit crusts falls on the depth interval of 1300-2000 m corresponding to the zone of the top parts of a guyot. On the slopes in the intervals of depths 2000-2900 m two-unit and one-unit crusts occur more often (Figure 5). The fact that separate units were formed at different depths and were fixed to definite areas of slopes demonstrates that each unit was formed under the influence of specific facial conditions.

As to the types of substratum, the main volume of three-unit crusts lies on reefal limestones and volcanogenic-sedimentary rocks widely developed on top parts of guyots. It is interesting to mention that on nannoforaminiferal limestones of Eocene only one-unit and two-unit sections of crusts lie. In the intervals of depths lower than 3000 m in the zone of petrified claystones only unit III occurs.

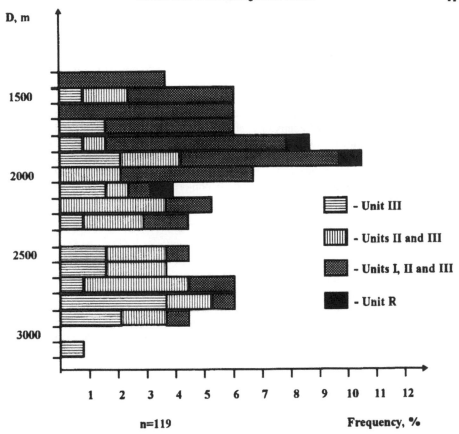

Figure 5. Frequency distribution of the crusts with different units according to the depth.

COMPOSITION OF FERROMANGANESE CRUSTS

Oceanic ferromanganese crusts are complex mineralization that contains major and trace elements: Mn, Fe, Ni, Co, Cu, Mo, Pt, REE, As, Hg, F, P. They refer to the rich cobalt geochemical type which is characterised by cobalt content more than 0,40 per cent, total Ni and Cu content less than 0,70 per cent and belong to the cobalt manganese type of ores [12].

Lower units differ greatly from upper units in the contents of main ore elements and mineral composition (Table 1). The ferruginous content increases from the base of the section to the upper part, which is accompanied by the fall of the value of the manganese modulus. In the same way the content of cobalt consecutively increases during which sharp changes are recorded in unit II where it is as high as 0,57 per cent compared with 0,36 per cent in Unit I. The content of phosphorus pentoxide in the upper units, on the contrary, sharply falls from 13,94 per cent and 9,52 per cent in Units R and I to 1,30 per cent and 1,27 per cent in Units II and III.

As to the mineral composition, unit R according to M.E. Melnikov is characterised by distinguishing features. It consists of asbolane and 5A mineral, vernadite and ferroxigite are determined here with some degree of conventionality, while for the upperlying units ferruginous vernadite and manganese ferroxigite are the main ore forming minerals, though in zone I-1 asbolane and 5A mineral are found in the form of admixtures. On the whole the units are characterised by the following mineral associations:

- unit I - vernadite - ferroxigite - apatite
- unit II - vernadite - ferroxigite - clayey minerals and feldspars
- unit III - vernadite - ferroxigite - quartz.

Non-ore component is traced visually in each unit. For units R and I with reference to upperlying units the increased contents of admixtures of carbonate-phosphate composition is remarkable, being represented in units R and zone I-1 by conformable veins and cross veins and in zone I-2 - by the material of pore filling. In upperlying units the sharp fall of carbonate component is noted. In unit II the material of ore filling is represented by suspension of predominantly clayey, more seldom carbonate-clayey composition. In unit III non-ore component is represented by thinly dispersed quartz [12].

AGE OF FERROMANGANESE CRUSTS

In the scanning electronic microscope against the background of micrite carbonate-phosphate mass which is present in the composition of ferromanganese crusts, a large quantity of fragments of calcareous nannofossils and discoasters has been found. Many of them have preserved their morphological features and have been identified to genus and in some cases to species. Thanks to these determinations it has become possible to single out 5 age groups of calcareous nannofossils and on this basis it is possible to make stratification of the main elements of crust section (Table 1).

Relics of unit R are the most ancient formations in ore section. Complexes of calcareous nannofossils revealed in conformable and cross veins suggest that: **the formation of unit R began in Late Cretaceous.** In syngenetical lenses and conformable veins (sample 15D183B) nannoplankton association, pertaining to the Upper Maastrichtian, has been revealed: *Micula mura* (Martini), *Watznaueria barnesae* (Black), *Zugodiscus* sp., *Lithraphidites quadratus* (Bramlette) (Plate 1). Domination in the complex of the species index *Micula mura* allows, to a certain degree of conventionality, to correlate it with the zone - of the same name - on Bukry's scale (70 - 65Ma) [3].

Post ore cross veins (sample 15D183B, 35D131) contain the complex of calcareous nannofossils in which the species of the late Paleocene - early Eocene dominate: *Discoaster multiradiatus* (Bramlette), *D. gemmeus* (Stradner), *D. diastypus* (Bramlette), *D. barbadiensis* (Tan Sin Hok), *D. lodoensis* (Bramlette et Riedel), *D. kuepperi* (Stradner), *Thorocosphaera operculata* (Bramlette et Riedel), *Chiasmolithus* sp. (Plate 1). With a certain degree of conventionality the revealed complex can be correlated with zones CP8 *Discoaster multiradiatus* - CP9 *Discoarter diastypus* (55.0 - 52.0 Ma) on H. Okada and D. Bukry's scale [13].

Carbonate - phosphate cement of breccias containing fragments analogous to unit R is characterised by calcareous nannofossils of Campanian-Maastrichtian: *Watznaueria barnesae* (Black), *Arkhangelskiella cymbiformis* (Vekishina), *Cribrosphaera ehrenbergii* (Arkhangelsky), *Biscutum constans* (Gorka), *Zygodiscus spiralis* (Bramlette et Martini).

The formation process of proto- unit R was evidently interrupted more than once. The evidence for it is the presence of fragments of crusts in Campanian-Maastrichtian breccias. Sporadic occurrence of unit R relics testifies to its destruction which took place not later than Paleocene according to the age of post ore cross veins. It is not excluded that ore fragments in Campanian-Maastrichtian

Table 1. Average content of the chemical elements in the different units of the ferromanganese crusts.

Age	Section of Fe-Mn crusts	Thick	Units	Minerals	Chemical composition, %					
					Mn	Fe	Co	Ni	Cu	P_2O_5
N_2-Q		0,1-4	III	Fe-Ver, Mn_Fer, Quarts	23,78	17,79	0,72	0,46	0,11	1,27
P_3^3-N_1^{1-2}		0-5	II	Fe-Ver, Mn-Fer, Clay minerals, Fillipsite, Feldspate	21,74	17,72	0,57	0,47	0,157	1,30
P_2^{2-3}		0-5	I-2	Fe-Ver, Mn-Fer	17,84	9,29	0,33	0,56	0,152	9,52
P_1^2-P_2^1		0-5,5	I-1	Fe-Ver, Mn-Fer, Apatite, Asbolan	20,17	11,54	0,38	0,48	0,131	7,45
K_2		0-8	R	Asbolan, 5A-mineral, Apatite, Fe-Ver, Mn-ferroxiagite	9,92	5,95	0,16	0,45	0,12	13,54

breccias and relics of unit R refer to different stages of ferromanganese oregenesis: the first - undoubtedly - to the Late Cretaceous, the second - possibly, to the late Paleocene.

Formation of unit - I took place in Eocene from the moment of disintegration of unit R. Carbonate - phosphate conformable veins in zone I - 1 (sample 35D131) contain the complex of calcareous nannofossils analogous to those revealed in post ore cross veins of unit R where the species index of the late Paleocene - early Eocene of zone CP8 *Discoaster multiradiatus* - CP9 *D. diastypus* also dominate(Plate 2).

In zone I - 2 carbonate - phosphate material of pore filling between ore substance includes species of the late Eocene. In a number of cases (sample 35D88) the complex of calcareous nannofossils is represented by the association corresponding to zone CP14 *Reticulofenestra umbilica* (45.0 - 42.0 Ma) on H. Okada and D. Bukry's scale [13]. The association consists of numerous species such as: *Reticulofenestra umbilica* (levin), *R. bisecta* (Gartner), *R. reticulata* (Persh-Nielsen), *Sphenolithus pseudoradians* (Bramlette et Wilcoxon) (Plate 3). In sample 36B22 the ore substance of zone I - 2 is characterised by the complex of calcareous nannofossils in which along with the above listed species are present the species corresponding to the middle Eocene: *Tranversopontis pulcher* (Deflaundre), *Cyclococcolithina formosa* (Wilcoxon), *Discoaster kuepperi* (Stradner), *Tharocosphaera* cf. *operculata* (Bramlette et Riedel).

Accumulation of ore substance of zone I-I undoubtedly took place in the early Eocene. The complex of calcareous nannofossils revealed in zone I-2 in crust of guyots Dalmorgeologia shows that accumulation of ore substance of zone I-2 might take place not only in the middle-late Eocene, as it was established for crusts of guyot IOAN, but also in Oligocene. There was possibly a break between periods of formation of zones I-1 and I-2.

The clayey suspension of unit II practically does not contain calcareous nannofossils with the exception of rare, badly preserved remnants. Good complexes are found only in the base of the unit where carbonate - phosphate macrointerspersions are noted. In samples 35D131 and 36D2 the base of the complex consists of the species characteristic of the late Oligocene - early Miocene: *Discoaster druggii* (Bramlette), *D. variabilis* (Martini) *D. deflandrei* (Bramlette et Riedel), *D. woodringi* (Bramlette et Riedel) (Plate 4). With a certain degree of conventionality the revealed complex can be correlated with subzones CN1b. *Discoaster deflandrei* - CN1c *Discoaster druggii* (23.0 - 18.0Ma) on H. Okada and D.Bukry's scale [13].

The time of formation of the lower half of unit II section can also be judged from the age of carbonate - phosphate material of cross veins (samples 15D3) sharply prevailing in unit I and interrupted in unit II. The complex of calcareous nannofossils revealed in cross veins allows their dating to the late and middle Miocene. Similar results have been obtained by H.M. Saidova for the crusts from guyot Ita - Maitaya, where she has discovered the complex of plankton foraminifers of the middle Miocene [2].

The sharp fall of the carbonate component in unit III makes it biostratigraphical reference difficult. A complex of calcareous nannofossils has been found in the upper part of unit III section in the only sample 36D2 in the carbonate inclusions. This complex is characterised by the following Pleistocene species: *Rhabdosphaera clavigera* (Murray), *Ceratolithus cristafus* (Kamptner), *Coccolithus pelagicus* (Wallich et Schiller), *Helicosphaera carteri* (Wallich et Kamptner), *Cyclococcolithina leptopora* (Murray et Blackman), *Umbilicosphaera mirabilis* (Lohman), *Gephyrocopsa oceanica* (Kamptner). A similar complex has been obtained by M.Y. Birulina from unit III in the crust of mount Jurioku. Paleomagnetic research of these crusts indicates that the formation of unit III took place in the Pliocene - Pleistocene epochs[11].

Calcareous nannofossils from cross and conformable veins carbonate-phosphate composition. Unit R. "Dalmorgeologia" guyot.

Plate 1. Late Paleocene - early Eocene (cross veins). 1 - *Discoaster multiradiatus* Bramlette, x4800. 2 - D. *cf .gemmeus* Stradner, x5400. 3, 4 - *D.cf.diastypus* Bramlette, x7200. Late Cretaceous (conformable veins). 5 - *Micula cf.mura* Martini, x10000. 6 - *Watznaueria barnesae* Black, x10000. Sample 15D183-B.

Calcareous nannofossils from conformable veins carbonate-phosphate composition.
Unit I, zone I-1. IOAN guyot.

Plate 2. Late Paleocene-early Eocene. 1, 4 - *Discoaster multiradiatus* Bramlette, x6000. 2 - *D.* cf. *gemmeus* Stradner, x1000. 3, 6 - *D. barbadiensis* Tan Sin Hok, x6000. 5 - *Thoracosphaera* cf. *operculata* Bramlette et Riedel, x3000. Sample 35D131.

Many facts suggest that formation of unit II began in the early Miocene. However the further history of unit II and III crusts formation remains not clear. The question of "in-situ" occurrence of Pleistocene complex in unit III is treated in different ways. It is not excluded that coccolites of this age are brought into the earlier originated crusts. Thus the upper temporal boundary for unit III formation could be estimated approximately.

The data presented suggest that the beginning of formation of **unit II dates back to the early Miocene**. It is not excluded that accumulation of **unit III took place through the Pliocene - Pleistocene epochs.**

DISCUSSION AND CONCLUSIONS

According to the structure, composition and age of crusts on guyots of the Magellan seamounts, three generations of ferromanganese crusts corresponding to three stages of formation can be distinguished[12].

Stage One - the Late Cretaceous (K_2). During this period formation of proto-unit R took place. In general terms the latter is characterised by the block composition of ore substance, the low content of main ore elements, the presence of large quantity of sedimentary material, the high degree of diagenetic transformations. Composition of manganese minerals differs sharply from the upperlying units, due to introduction of a certain part of the substance from hydrotherms. It is not improbable that the process of ore substance accumulation in the Late Cretaceous was of intermittent character and was periodically interrupted. The most favourable conditions for crust forming appeared in Campanian - Maastrichtian, when on the majority of guyots the relatively stable bathyal regime of sedimentary accumulation was set up. The sporadic occurrence of proto-unit R relics evidenced its disintegration and practically complete destruction.

Stage Two - Eocene (P_2^{1-3}) - embraces the time span of formation of unit I. The composition of ore substance is close to that of the present day, with some features similar to the composition of proto-unit R. Ore substance accumulation of unit I took place on gently-sloping relief in top parts zones, more rarely on slopes. Sedimentation rate of zone I-1 was relatively low (syngenetic structures are thinly-parallel laminar), zone I-2 was formed quicker, as evidenced by large columnar structures. A number of facts suggests local interruptions in the growth of unit I. One of them is the absence of zone I-2 from the section, underlined in a number of cases by a sharp contact with upperlying units.

Stage Tree - Miocene - Pleistocene (N_1^1-Q). During this period formation of units II and III took place which in their structure-textural features and chemical composition differ greatly from units R and I. Unit II and III are affected by diagenetic processes to a lesser degree. The content of cobalt increases sharply, the content of iron and manganese also increases, whereas the content of phosphorus pentoxide sharply falls. Non-ore material is represented by clayey and carbonate-clayey inclusions. The area of occurrence of generation three crusts considerably expands. Ore substance accumulation takes place not only in the top parts zones, but on the slopes of guyots. The origin of unit II evidently takes place at the end of Oligocene-beginning of Miocene. Giant-columnar structure suggests its relatively quick forming. The beginning of unit III formation should be possibly referred to the beginning of Pliocene. Its structure and composition reveal that the formation took place under conditions somewhat different from, but fairly close to those of the forming of unit II. It is not improbable that units II and III can be referred to different generations of substance, but without decisive arguments we, following P. Halbach and co-authors, refer these

Calcareous nannofossils from ferromanganese matrix. Unit I zone I-2. IOAN guyot.

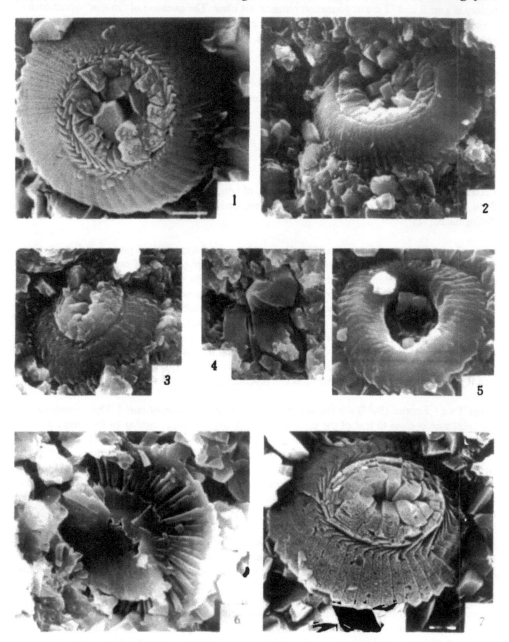

Plate 3. Middle-early Eocene. 1 - *Reticulofenestra reticulata* Perch-Nielsen, x6000. 2 - R. *umbilica* Gartner, x6000. 3, 7 - *R.bicecta* (Hay, Mohler et Wade)Roth, x6000. 4 - *Sphenolithus pseudoradians* Bramlette et Wilcoxon, x6000. 5 - aff. *Cyclicargolithus floridanus* (Roth et Hay) Bykry, x10000. 6 - aff. *Cribrocentrum martini* Hay et Tawe, x6000. Sample 35D88.

Calcareous nannofossils from ferromanganese matrix. Unit II. IOAN guyot.

Plate 4. Early Miocene. 1 - *Discoaster adamanteus* Bramlette et Wilcoxon, x6000. 2 - *D. Woodringii* Bramlette et Riedel, x8000. 3, 4 - *D. deflandrei* Bramlette et Riedel, x6000. 5 - *D. aulacos* Gartner, x5400. 6 - *Coccolithus miopelagicus* Bukry, x4000. Sample 35D131.

units to general "younger generation" [7, 12]. Its formation has been evidently continuing up to now in areas with favourable facies conditions.

The accumulation process of ferromanganese crusts has been lasting from the Late Cretaceous to Pleistocene inclusive, but punctuated by numerous interruptions. Two of them are regional. The first one is between generations one and two of crusts on the boundary Cretaceous-Paleocene. The second is between generations two and three on the boundary Eocene-Oligocene. Other interruptions are apparently connected with local changes of facies conditions of ore substance accumulation.

The bedding conditions of ferromanganese crusts, their stratified structure, the unit - to -unit change of their structure-textural features, chemical and mineral composition, degree of diagenetic transformations support the sedimentary origin of the ferromanganese ores. On the ground of the collected data and the theory of sedimentary origin of oceanic ferromanganese crusts we come to the conclusion that the history of ferromanganese crusts formation is correlated with the evolution of contribution of biogenic calcium carbonate to the ocean floor and its subsequent dissolution.

- In the Late Cretaceous a great evolutionary expansion of calcareous microplankton began, giving rise to wide distribution of planktonogenic carbonate sediments of the age in question [10].
- On the boundary of the Early and Late Cretaceous the atoll stage of guyot development was interrupted. Ruin of biocenosis of reef structures was fixed. On the slopes of submerging guyots accumulation of nannoforaminifer sediments took place.
- The beginning of ferromanganese crusts formation of the early generation is dated to the Late Cretaceous.
- On the boundary the Late Cretaceous - Paleocene mass extinction of species of plankton foraminifer and calcareous nannoplankton took place, which led to the sharp reduction of carbonate accumulation in the Pacific. On this boundary regional break in carbonate accumulation is marked. During this period interruption in ore substance accumulation occurs.
- By the middle Paleocene a variety of species of calcareous microplankton was gradually restored, it continued to grow and reached its maximum in the middle Eocene[10].
- In Eocene carbonate accumulation on the greater part of the area of the Pacific became continuous, though slow.
- In Eocene the majority of the Magellan seamounts guyots stabilised on the depth approximate to those of the present day. On the top parts and slopes nannoforaminifer sediments accumulated.
- At the Paleocene - Eocene boundary under favourable conditions formation of generation two of ferromanganese crusts took place.
- At the Eocene - Oligocene boundary a considerable development of ice at high latitudes and global cold spell occurred, which caused global crisis of biota. Intensive extinction of plankton microfossils and reduction of variety of   species occurred. Oligocene sediments reflected low biological   productivity[10].
- During the early Oligocene regional break in carbonate accumulation is marked. It apparently correlates with the break in ore substance accumulation that took place between stages two and three of ferromanganese crusts formation.
- In Miocene the Earth climate entered the ice stage. By 15 million years the sizes of the Antarctic ice shield considerably increased. The increased climate contrast between the higher and lower latitudes led to the intensified circulation of oceanic waters, as well as to the increase of nutrient supply from surface and subsurface waters to the area of preferential development of plankton life[10].
- In response to new paleooceanologic conditions evolutionary radiation   resumed. At the beginning of the middle Miocene biological processes increased   more sharply.

- The early Miocene was the time when under other favourable conditions forming of ferromanganese crusts of generation three began, which constituted the main volume of ore deposits of the Magellan seamounts guyots.

In this article one of the possible version of interpretation on the received data is presented. Further research will allow to specify and supplement crust stratification of the West Pacific, and research of crust development in other areas will considerably replenish notions on the development of the oceanic ferromanganese ore genesis.

Acknowledgements

The author expresses her gratitude to geologists of Kamchatka geological party of the Joint-Stok company Dalmorgeologia, who carried out their research in the Central Pacific from 1983 to 1993; M.M. Zadornov, who supervised this research; Er.B. Nevretdinov and M.E. Melnikov for his help in setting the tasks, L.A. Kartseva for her help in the work with scanning microscope and I.V.Belokopitov and O.A.Belokopitova for technical assistance in computer typesetting of this article.

REFERENCES

1. Baturin G.N. Geochemistry of ferromanganese nodules of ocean, M.: Nauka, 1986, 229p. (in Russian).
2. Bogdanov Ju.A., Sorokhtin O.G., Sonenshine L.P. et al.. Ferromanganese crusts and nodules of the Pacific Ocean seamounts, M.: Nauka, 1990, 288p. (in Russian).
3. Bukry D. Biostratigraphy of Cenozoic marine sediment by calcareous nannofossils, *Micropaleontology*, 1978, vol. 24, № 1, p. 44-60.
4. Bukry D. Coccolith stratigraphy, eastern equatorial Pacific, Leg 16, Deep Sea Drilling Project. *Initial Reports of DSDP*, 1973, V.16, p.653 .
5. Cronen D., Submarine mineral deposits. M.: Mir, 1983 (in Russian).
6. Glasby G.P., Andrews J.E., Manganese crust and nodules from the Hawaiian Ridge. *Pacific Science* , 1977, V. 31, 4, p.363-379.
7. Halbach P. and D. Puteanus, The influence of the carbonate dissolution rate on the growth and composition of Co-rich ferromanganese crusts from central Pacific seamount areas. *Earth Planet. Sci. Lett.*, 68, 73-87, 1984.
8. Hein J.R., J-K. Kang, M.S. Schultz et al., Geological, geochemical, geophysical, and oceanographic data and interpretations of seamounts and Co-rich ferromanganese crusts from the Marshall Islands, KORDI-USGS R. V. Farnella cruise F10-89-CP, U.S. Geol. Surv. *Open File Rep.*, 90-407, 264 pp., 1990a.
9. Hein J.R., F. T. Manheim, W.S. Schvab, A.S. Davis, C.L. Daniel, R. M. Bouse and I.A. Morgenson, Geologic and geochemical data for seamounts and associated ferromanganese crusts in and near the Hawaiian, Johnston Island, and Palmyra Islands Exclusive economic Zones. U. S. Geol. Surv. *Open File Rep.*, 85-295, 129 pp., 1985.
10. Kennet J.P., Marine Geology. Vol 1. M.: Mir, 1987 (in Russian).
11. Linkova T.I., Ivanov Ju.Ju., Magnitostratigraphic study of ferromanganese crusts from Central Pacific . *Tikhookeanskaya Geologia*, 1992, №2, p.p.3-11 (in Russian).
12. Melnikov M.E. Pulayeva I.A., Ferromanganese crusts of Markus-Wake rise and the Magellan seamounts of the Pacific Ocean. *Tikhooceanskaya Geologia*, 1994, №4, p.p. 13-27 (in Russian).
13. Okada H., Bukry D., Supplementary modification and introduction of code numbers to the lowlatitude coccolith biostratigraphic zonation (Bukry, 1973, 1975). *Mar. Micropaleontol.*, 1980, V. 5, 3, p.321-325.

Proc. 30th Int'l. Geol. Congr., Vol. 13, pp. 129-144
Wang & Berggren (Eds)
©VSP 1997

Paleo-Ocean Events and Mineralization in the Pacific Ocean

DONGYU XU

Institute of Marine Geology, MGMR, P.O.Box 18 Qingdao,266071, China

Abstract

Deep-sea polymetallic nodules, Co-rich manganese crusts and seamount phosphorite are sedimentary mineral resources formed in certain geologic periods and in special geologic environmental conditions of Cenozoic. Results of Deep Sea Drilling Project (DSDP), Ocean Drilling Project (ODP) and Clarion-Clipperton (CC) core studies and correlations of textures, structures, minerals and geochemistry between the nodules, crusts and phosphorite have shown that paleoocean ographic events and environments such as Antarctic Bottom Water, upwelling, bioproductivity and hiatuses are controlling the formations and distributions of the above 3 types of mineral resources.

Key words: Polymetallic nodule, Co-rich manganese crust, Paleo-ocean

INTRODUCTION

Polymetallic nodules and Co-rich manganese crusts are the most interesting potential mineral resources. At present, many countries have turned regional reconnaissance survey of the nodules into detailed prospectings or explorations in smaller areas, and also fundamental scientific researches for the crusts are turned into regional resource investigations. Therefore, the knowledge of formation environments and distributions of polymetallic nodules and crusts is both of theoretic and practical significence. Halbach et al. [7, 8, 10, 11] , Hein [13] , Andreev [35] Skornyakova [37] and Cronan [6] et al. approached the formation environments and distribution regularities of polymetallic nodules and Co-rich manganese crusts in the aspects of sedimentology, paleooceanography and geochemistry. The author of this paper tries to use the view of global tectonics and paleo-ocean evolution as the tool, on the basis of China's investigations and studies on polymetallic nodules and crusts in the Northeast Pacific Ocean(Fig 1, Tab.1), to prove into growth history and distribution of the deep-sea ferromanganese sediments.

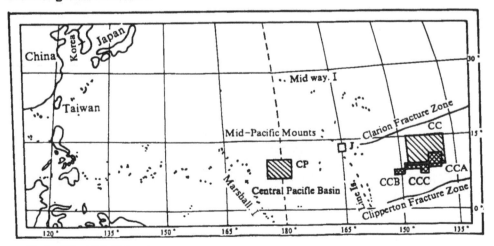

Fig.1 Map of studied area

Tab.1 List of samples studied

station	latidute	longitude	Water depth	Topography	sediments	type of nodules
CP2	07 ° 02′ N	176 ° 14′ E	5369	hill		M(SP)
CP22	07 ° 30.66′ N	177 ° 1.64′ E	5356			
CP23	08 ° 17.54′ N	177 ° 0.82′ E	5960			
CP25	10 ° 44′ N	179 ° 31′ E	6090	plain	S.C	S(SP)
CP30	07 ° 02′ N	176 ° 31′ E	6303	plain	S.C	
CC3	13 ° 59.84′ N	144 ° 0.09′ W	5390		S.O	
CC48	09 ° 47′ N	144 ° 00′ W	5022	hill	S.C	S(S)
CC61	11 ° 15.88′ N	142 ° 59.98′ W	5220			
CCA5	9 ° 2.27′ N	141 ° 14.37′ W	5198		S.C	
CCA15	10 ° 15′ N	139 ° 29′ W	4918	hill	S.C	S(T)
CCA56	10 ° 14.86′ N	139 ° 14.97′ W	4771		CSO	
CCA57	10 ° 00′ N	139 ° 14′ W	5015	mount	C.O	S(T)
CCA72	09 ° 10′ N	140 ° 13′ W	5077	plain	S.C	M(E)
CCA121	10 ° 15′ N	140 ° 50′ W	5009	hill	S.C	
CCB21	08 ° 46′ N	147 ° 00′ W	5250	plain		L(C)
CCC9	08 ° 00′ N	142 ° 58′ W	5058	plain	S.C	
CCC30	08 ° 41′ N	141 ° 59′ W	5073	plain	S.C	L(C)
5340	09 ° 44′ N	151 ° 37′ W	5172		S.C	
5352	09 ° 37′ N	151 ° 59′ W	4867	mountain	S.O	M(E)
5381	09 ° 22′ N	152 ° 14′ W	5217	hill	S.C	
5397	09 ° 15′ 00 ″ N	151 ° 52′ 31 ″ W	5199		S.O	
5342	09 ° 37′ 30 ″ N	153 ° 22′ 32 ″ W	5290	S.C		

Sediments: C.O—Calcareous ooze, C.S.O—Calcareous—silieous ooze,
S.C—Siliceous clay, S.O—Siliceous ooze
type′ of nodules: M(SP), indicate the size and forme of module, that is
M—mediuon size   L—large size   S—small size
(C)—Cauliflower—like,   [E]—Ellipsoidal
(SP)—Spherical Polynucleated   (S)—spherical   (T)—Tabular

DISTRIBUTIONS OF POLYMETALLIC NODULES AND Co-RICH MANGANESE CRUSTS IN THE PACIFIC OCEAN

At the tops and on the slopes of the Pacific seamounts or ridges formed before

Eocene, very commonly are found Co–rich and Ni–and Cu–poor crusts, and manganese minerals are mainly $\delta-MnO_2$. The total rare earth element content is very high, generally higher than 1000ppm. The matrix of the crusts is mainly volcanic rock or debris, and the crust cores are mainly debris, clay or bioclast. At the tops of the seamounts, there are also phosphorite nodules, and in the crustified beds usually exist phosphate sediments. The associated sediments are mainly calcareous ooze, indicating that Co–rich crusts and phosphorite are formed above CCD (Carbonate Compensation Depth).

In abyssal plains and abyssal hills (under CCD) of Eocene–Miocene origin, nodules are hydrogenic and diagenetic, rich in Mn, Ni and Cu and poor in Co and Fe. Manganese minerals are dominated by todorokite and $\delta-MnO_2$. Rare earth content is relatively low, usually less than 1000ppm. And associated sediments are mainly siliceous ooze and siliceous clay. In older abyssal plains and hills (formed before Eocene), nodules are mainly hydrogenic, rich in Fe and poor in Cu and Ni. Manganese minerals are dominated by $\delta-MnO_2$, and associated sediments are mainly siliceous clay, abyssal clay or zeolitic clay (Fig.2) [38,40,43]. Thus it can be seen that distributions and geochemistry of nodules and crusts are obviously influenced and controlled by tectonic movements, volcanic activity, topography and sediments (Fig.2 and 3).

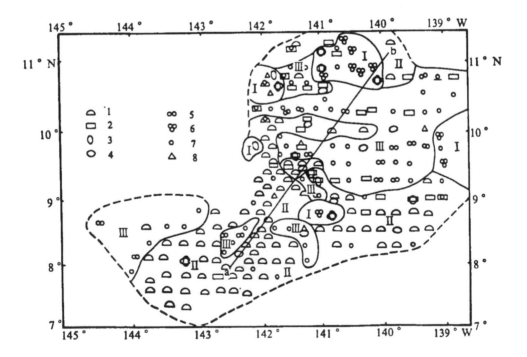

Fig.2 Distribution of morphological types of polymetallic nodules in Clarion—Clipperton area
I Spheroidal—ellipsoidal II Cauliflower—disc form III Bayberry—tabular
1.Cauliflower 2.Tabular 3.Ellipsoidal 4.Discoidal 5.Ellipsoidal and its intergrowth 6.Ellipsoidal and its growth of stock 7.Bayberry form 8.Clastic

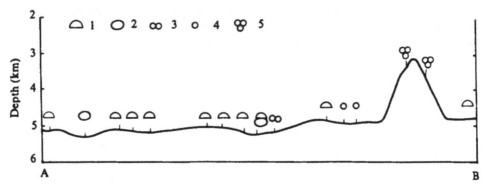

Fig.3 Morphological types of nodules in different geomorphologic units
1. Cauliflower   2. Discoidal   3. Growth of stock   4. Bayberry form   5. Ellipsoidal and its growth of stock

A number of buried nodules and crusts were reported in the cores of DSDP and ODP in the Pacific and Atlantic Ocean [30] . Shallow buried nodules were found in cores CP25 and CC48 in the Central Pacific Ocean (Fig.4) [38,40,43] . Most of these buried nodules occur at a sedimentary hiatus, which indicates that the condition of formation of nodules is the environment of hiatus or very slow sedimentations [40,43] .

GROWTH HISTORY OF NODULES AND CRUSTS

Growth history of noudles

Both hydrogenic and diagentic nodules have concentric laminations or dendritic structures. And between different growth layers of nodules, usually appears a parallel or oblique growth hiatus, indicating a discontinuous rather than a continuous growth, characteristic of a variable growth environment(Fig.5) [43] .

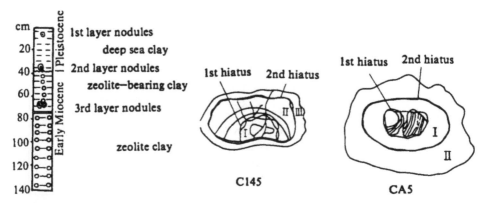

Fig.4 Sedimentary section of CP25 Core [43]   **Fig.5** Hiatus of polymetallic nodules [43]

According to the growth hiatus of nodules, several structural layers or growth layers can be recognized, and then isotopic dating or paleontologic analysis method is used to determine the age (growth period) of each growth layer. Based on the analysis of Han Changfu et al.(1994) [43] , nodules from CC area of the East Pacific have 3 growth periods: the first growth period is from late Eocene to middle early Miocene, the second from early Middle

Miocene to the end of late Miocene and the third from Pliocene to Pleistocene(Tab.2). According to K—Ar dating, nodules from CC3 and CC61 (CC area in the East Pacific Ocean) and from CP22 (in the Central Pacific) have their first growth layers respectively dated to be 32.6 Ma, 32.7Ma and 26.9Ma B.P., roughly corresponding to middle and late Oligocene. The third growth layers of nodules from CP23 (in the Central Pacific) are dated to be 4.6Ma B.P., corresponding to Early Pliocene. According to the [10]Be dating, the second growth periods of nodules from CCC30 and CCB21 (CC area between Clipperton Fracture zone and Clarion Fracture zone) and from CP2 (the Central Pacific) are all during 6Ma B.P., that is, late late Miocene, and their third growth periods are during 4—6Ma B.P. (early Pliocene) [42] .

Table.2 Fossils occurring during different growth periods of nodules

Growth period	Fossils
First growh period E_3–M_1^1	1. *Ericsonia* cf. *robusta, Coccolithus* sp. *Triquetrohabdulus carinatus, Discoaster cubensis, D.druggii, D.deflondrei, Coronocyclus* sp. 2. *Actinoptychus undulata, F.octoplicatus,* A.Cl. 3. *Theocyrtis onnosa, Lithomitra lineata, Dictyoprora mongolfieri*
Second growth period	1. *Discoaster aulakes, Coccolithus pelagicus, Sphenolithus moriformis* 3. *Dorcadospyris simplax, D.dentata, Dendrospyris didiceros*
Third growth period	1. *Gephyrocopsa oceanica, Cyclocolithus leptoporus* 2. *Thalassiosira oestrupii, Pseudoeunotia doliolus, Hemidiscus cueiformis, Nitzschia marine* 3. *Thyrsocyrtis tetracantha, Lirlospyris stauropora*

1.Calcareous nannofossil
2.Diatom
3.Radiolaria

Growth history of Co-rich manganese crusts

Halabach et al. [11] thought Co—rich manganese crusts to have two growth layers, one is younger than 10 Ma, and the other is during 40— 38Ma B.P., perhaps from middle Eocene to early Oligocene. According to Ikari and Nishimura [15] , the geologic time for Co—rich crusts on the Tenpo Seamount of Nishi—Shichito Island Ridge to have begun growing was late Miocene, which is inferred from the birth of the Tenpo Seamount, on which the crusts depended to grow. This seamount, a volcanic is land, was formed 15 —16Ma B.P. or even earlier.

CENOZOIC TECTONIC AND PALEOOCEANOGRAPHIC EVENTS IN THE PACIFIC OCEAN

Systematic researches of DSDP and ODP cores in combination with various geophysical results, have reconstructed the environmentary evolution and paleoceanographic events since Mesozoic such as lithospheric plate movements, oceanic distribution patterns and currents. Due to seafloor spreading and plate movements since the Late Cretaceous, a series of valcanos, ridges and fracture zones came into being on the Pacific floor such as the Music Seamount, the Lateyev Seamount, the Line Islands Ridge, the Marshall—Gilbert Ridge, the Molokai Fracture Zone, the Murray Fracture

Zone, the Clarion Fracture Zone and the Clipperton Fracture Zone. In Eocene, seafloor movements changed in direction and velocity, which made some weak zones and volcanos more active. A great amount of Eocene basalt occurred in the Line Islands Ridge, Marshall— Gilbert Ridge and some other places of Cretaceous origin. And during the Miocene, the Pacific Ocean floor had a series of spreading centers formed and seafloor hydrothermal activity became much more intense. For example, hydrothermal activity is still active on the East Pacific Rise.

The seafloor tectonic events have not only changed topography but also environmentary factors such as oceanic currents (Table 3). It is generally known that Australian continent drifted northwards away from the Antarctic continent in late Paleocene and early Eocene (55— 53Ma B.P.), and ocean was formed between the two continents [34], and circum—Antarctic circulation and Antarctic ice began to occur [22]. At the end of Eocene, the Tasman Rise sank, which linked up the Indian Ocean with the Pacific [22,23], and circum—Antarctic shallow current and Antarctic Bottom Water began to form, climate became colder [34] and vast hiatus appeared [25]. In early Oligocene (38—35Ma B.P.), the South Pacific Ocean was linked to the Atlantic by the Drake Passage and then the ocean was thoroughly open through the Drake Passage (31 —23 [1], the circum—Antarctic circulation completely formed, New Zealand Boundary Current occurred, upwelling developed [22], and temperature gradient between regions was increased [21]. In Oligocene, the developed circum—Antarctic circulation kept warm subtropical circulation away from cold sub—Antarctic circulation and thus the Antarctic had no heat supply any more. As a result, glaciation increased, and Antarctic glacial sheet and sea ice spread in Miocene. Thein crease of glaciation and the development of circulation resulted in the expansion of Antarctic water mass and the formation of Antarctic convergence current in early Miocene when siliceous bioproductivity increased greatly and siliceous sediments spread northwards [3,22]. In middle Miocene, most of the Antarctic glacial sheet came into being [28,29], which caused an intense activity of the Antarctic Bottom Water and equatorial countercurrent, and hiatus appeared widely in the Equatorial Pacific [17,18,19,20,32]. In late Miocene, the Mediterranean Sea was enclosed because of the uplift of Gibraltar [14], global climate turned colder(6.5—5.0Ma B.P.) [24], the Antarctic glacial sheet spread, the Bottom Water was completely formed and intensified, the Pacific circulation also became intense, high—latitudinal and equatorial Pacific upwelling currents were well developed [26,33], and a large amount of phosphate was deposited in neritic areas [4]. The intensified glaciation at the end of the Miocene resulted in a large drop of the sea level [31], which induced the Messinian Event (6.2—5Ma B.P.) that dried the Mediterranean Sea [2] and re sulted in the salinity crisis [5]. During the Pliocene, the Mid—American passage between the South and North America was closed (3.1—3.4Ma B.P.), and the Pacific Ocean was cut off from the Atlantic [16]. The Pacific circulation system and Antarctic and equatorial upwellings became much stronger, and siliceous bioproductiviy increased.

In order to know paleoocean evolution in the Central Pacific Ocean, we did magnetostratigraphic, lithostratigraphic, biostratigraphic and isotope geological analyses of cores recovered from station Cp 30 in the Central Pacific Basin and from stations CC48, CCA121 and CCP in CC area of the Northeast Pacific (Fig.6), and $\delta^{13}C$ method was used for a roughl estimation of bioproductivity in different Miocene periods [14].

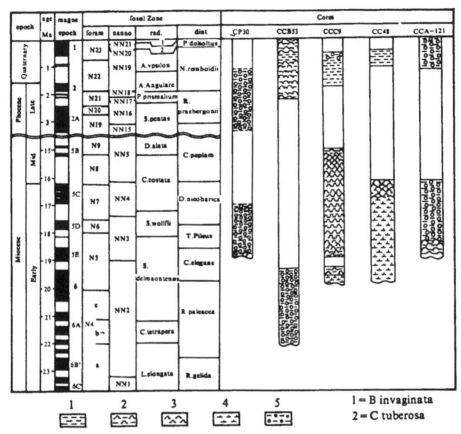

1 = B invaginata

2 = C tuberosa

Fig.6 Stratigraphic Correlation of Late Cenozoic in the Central Pacific [41]
1. deep sea clay;   2. siliceous clay;   3. siliceous ooze; 4.   calcareous ooze;   5. zeolitic clay

Isotopic and paleontologic analyses for CP30 and CC48 show that CC area of the Northeast Pacific Ocean saw two $\delta^{18}O$—rich events in early early Miocene (22Ma—20Ma B.P.)(Fig.7), and in equatorial and tropical biologic assemblages have found cold—water species such as *Coscinodiscus marginatus* (diatom), *Coccolithus pelogicus* (nannofossil) and *Epsistominella exiqua* (benthic foraminifera), and meanwhile two $CaCO_3$ dissolution events took place(Fig.8). Thus it can be inferred that as early as in early Miocene, the early Antarctic Bottom Water already entered CC area and hiatus occurred in some localities. During middle Miocene (20—18Ma B.P.), there were three $\delta^{18}O$—rich events lasting from 20.33 to 19.88Ma B.P., from 19.55 to 19.26Ma B.P. and from 18.72 to 17.81Ma B.P., respectively. As compared with the events before 20Ma B.P., the $\delta^{18}O$ was getting richness, indicating that the climate was becoming colder. The two $CaCO_3$—dissolution events of 20—19.12Ma B.P. and 19.73—19.00Ma B.P. caused two periods of hiatus, roughly corresponding to NH1a determined by Keller and Barron [20]. During the periods, bioproductivity was less than 200mC / m²a. In late early Miocene (18—16Ma B.P.),there was a wide hiatus corresponding to NH1b in CC area, and most of the seafloor was below CCD, which is proven by CC48, CCA121 and CCB53 cores, where *Calocycletta costata* Zone was absent in late early Miocene siliceous ooze and siliceous clay beds. In middle Miocene, there was a wide regional

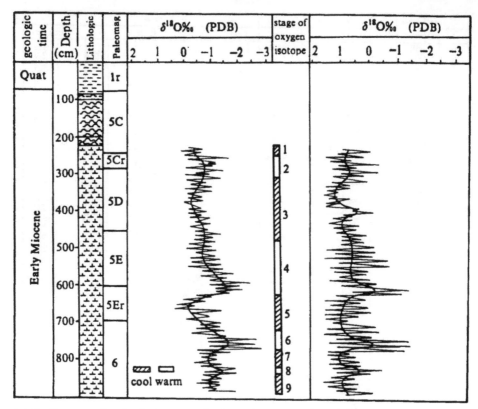

Fig.7 Oxygen and carbon isotope curve for the geological age of core CC48

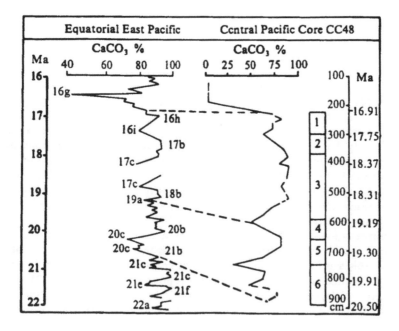

Fig.8 CaCO₃ cycle in core CC48 and in equatorial East Pacific Ocean [45]

hiatus, the Antarctic Bottom Water was very active, CCD rose, and mainly siliceous sediments such as radiolarian ooze and clay were deposited. The hiatus corresponds to NH2 and was confirmed by the fact that no Miocene standard fossils are found in core samples except in some surficial sediments where only a few typical fossils are seen. In late Miocene, global cold climate event affected the study area, the Antarctic Bottom Water became more active, and a regional hiatus occurred in most of the area, roughly corresponding to NH5 and NH6. In Pliocene, the study area saw the continuity of hiatus from the end of Miocene to early Pliocene, which was proven by the fact that no Early Pliocene standard fossils are found in the seafloor sediments, but many cold—water species such as *Coscinodiscus marginatus* in the sediments, indicating well—developed upwellings and high siliceous bioproductivity.

COMPARISON BETWEEN THE GROWTH HISTORY OF NODULES AND CRUSTS AND EVOLUTIONARY HISTORY OF THE PACIFIC OCEAN

Studies show that growth and distribution of polymetallic nodules and CO—rich manganese crusts are controlled by many oceanic and geologic factors and are closely related to certain paleooceanographic events.

As mentioned above, submarine topography such as seamount, abyssal plain and hills created by tectonic movements have influenced and controlled the types and compositions of ferromanganese sediments. The associated volcanism has supplied ore—forming elements (such as Mn and Fe) and nuclei or matrix for the growth of polymetallic nodules and Co—rich manganese crusts. Isotopic analysis nodules from the CC area shows that Pb isotope is relatively in depletion, while $^{206}Pb / ^{204}Pb$ value is between 18.563 and 18.745, with 18.693 as the average, indicating that the nodules were once influenced by hydrothermal activity (Sun Zhiguo,1995). It is known from statistics of Levitan and Lisitsin [36] that frequency of volcanic ash occurring in DSDP cores has a tendency to increase since Paleocene. the frequency is the highest in Middle and late Miocene and in Pleistocene and it can be up to 15%. Early Miocene— Pliocene and Quaternary sediments at stations CCC9, CCA121 (CC area) and CP30 (the Central Pacific) are mainly made up of zeolitic clay and zeolite—bearing clay, 10—20% of which are altered products of volcanic ash such as phillipsite and montmorillonite [41]. Almost all of the studied nodules contain phillipsite. For examples, nodules from CCA72, CCA15 and CCB56 (all in the CC area) have phillipsite up to 2.9%, 8.3% and 6.8%, respectively, where as the percentage in CCA57 nodules can be as high as 31.6%. Thus it can be inferred that growth and distribution of nodules are influenced by volcanism and volcanic products [40].

From shapes, surface structures and distribution characteristics of nodules and crusts, we know that they grow up after a long time uncovered on seafloor. Nodules or crusts on seamounts and hills are mostly smoothly spheroidal, in large or medium size, often rolling about under current action. If each of their growth layers is 0.1—1cm thick and the growth rate is 1mm / Ma, a nodule would take 5—10 Ma to grow to the present size. The semi—buried nodules in abyssal plain and hill areas have a smooth top and rough bottom, and these large—or medium—sized nodules have grown up at seawater—sediment interfaces. The thickness of their growth layer ranges generally from 0.1 to 1cm. If calculated at 2mm / Ma, a growth rate determined with isotopic method, the growth layer at least takes 2.5—5Ma to complete.

In the studied nodules, young fossils are often mixed with older ones and cold—water species are also found. For example, in growth layers of nodules, *Dictyoprora mongol fieri* (Eocene radiolaria) often mixes up with Miocene and even Pleistocene fossils, and is also associated with *Coscinodiscus marginatus*(cold–water species diatom) and *Coccolithus pelagicus* (cold–water species calcareous nannofossil), which indicate that nodules grow in the conditions of erosion of cold–water mass, suspension transportation and then redeposition [41,42,43,44].Mineral chemical analysis shows that minerals in nodule or crust layers which directly contact sea water are mainly made up of δ–MnO_2(quadravalent Mn) and rich in Fe, Co and Pb, and rare earth content is high (>1000ppm); while at the bottom of the nodule contacting with sediments, there is rich todorokite consisting of both quadravalent and bivalent Mn besides δ–MnO_2, and rare earth content is relatively low (<1000ppm), which also indicates that nodules are long uncovered in oxidation conditions [40].

We can infer from the above that nodules being long uncovered on the seafloor is made possible by hiatus and erosion of bottom water, rather than by any other factors such as the socalled "being turned up by organisms" or the pushing up of static pressure. This can be confirmed by the fact that underwater photographs and television at nodule–producing areas haves't found enough benthic organisms to turn up nodules.

Correlation analysis shows that the growth periods of nodules and crusts coincide with hiatus and cold periods of the Pacific Ocean, which is shown in Fig.9, Fig.10 and Table 3. Figure 9 is on the basis of the study results from CC48 core in CC area of the Northeast Pacific. We can see from the figure that the 3 growth periods of the nodules respectively correspond to hiatus in early Early Miocene (20.5—19.41Ma B.P. and 19.41 —19.21Ma B.P.), Middle Miocene (16.78—16.2Ma B.P.) and Late Miocene (1.61 Ma B.P.) [40,43]

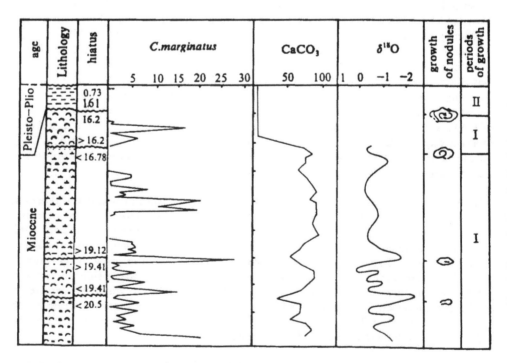

Fig.9   Comparison between paleooceanographic evolution history and nodule growth periods

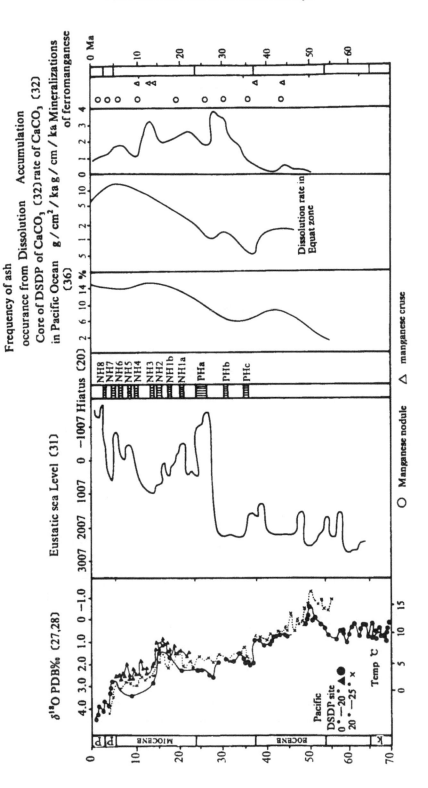

Fig. 10   Environmental evolution of the Pacific Ocean

Table 3. Evolution of ancient ocean and sedimentary mineral resources in the Pacific Ocean

Geologic time			Geological events	Paleooceangraphic events	Paleoclimatic events	Hiatus	Biological events	Sedimentation	CCD	Sedimentary minerals		
Stage	Age (Ma)									Nodule	Crust	Phosphorite
Holocene										0.5Ma		Coast of Peru West coast of Africa
Pleisto.	2		Closure of isthmus of Panama (3.1–3.6Ma) (3.1–3.5Ma)		Formation of Arctic Glacier and ice sheet				Deep	3.5Ma		California
Pliocene	5	Late		Development of Antarctic and equatorial up welling		NH7 (3.2–4.8Ma)	Increase of Siliceous bioproductivity	Formation of recent Oceanic sedimentary Pattern				
		Early										
Miocene	11	Late	Closure of Mediterranean rising of Gibraltar	Further development of Antarctic Bottom Water and climate Falling of sea level	Cooling event (11–9Ma)	NH6(7.6–6.2Ma) NH5(10–9Ma)	Expansion of siliceous Biogeographic zone	Increase of sediment tation rate at high bioproductivity zone	Shallow	5.4–4.8Ma 7.9Ma 10Ma	10Ma	California
		Middle		Formation of equatorial undercurrent; strengthening of equatorial countercurrent; development of upwelling at high latitude in North Pacific Ocean			NH4 (12–11Ma) NH3 (13.5–12.5Ma) NH2 (16–15Ma)	Expansion of planktonic biogeographic zone Disappearance of Paleogene planktonic foraminifers Renewing of planktonic fauna			12Ma 14Ma	12Ma 14Ma	Shore of California
	16	Early	Subsidence of Ireland—Faeroe ridge Collision of Europe and Africa Continent	Closure of Tethys Cutoff of Mid–American Deep Water Passage		NH1 (20–18Ma)	Development of radiolaric; Increase of siliceous bioproductivity; Expansion of calcareous nannoplankton biogeographic zone; Increase of planktonic foraminiferal diversity					
Oligocene	23	Late	Opening of Drake Passage (30–23.5Ma)	Formation of circum—Antarctic current and New Zealand Boundary current		pH (23.0–22.5Ma)			4500m (30Ma)	2.6Ma		
	30	Early		Formation of Antarctic sea ice, beginning to form Antarctic Bottom Water								
Eocene	37	L	Subsidence of south Tasman Sea	Connection of Indian and Pacific Oceans	Cooling event at the end of Eocene	Extensive hiatus			3200m	3.8Ma	3.8Ma	Pacific seamounts
	40	M										
	52	E	Separating and north ward movement of Australia from Antarctic Continent	Circus—antarctic current and ice began to form						4.5Ma	4.5Ma	Pacific seamounts
Paleocene	58											

Figure.10 and Table 3 are a composite correlation on the basis of some scientists' study results. It can be seen from the Figure 8 and Table 7 that the main growth periods of nodules and crusts correspond with important paleooceanographic events since Eocene such as global climate becoming cold, a large drop of the sea level, rapid dissolution of $CaCO_3$, hiatus development and frequent submarine volcanic activities. Table 3 also reflects the changes in ocean and continent distributions caused by global plate movements, which have resulted in the formation and development of oceanic current systems and changes in paleobiogeography and bioproductivity. The occurrence of the Antarctic Bottom Water and circum—Antarctic circulation is the most important paleooceanographic event to have changed Late Cenozoic climate and ocean environment. The activity of the cold and oxygen—rich bottom water has influenced or changed oceanic biogeography, bioproductivity and dissolution and precipitation of $CaCO_3$ and produced depositional hiatuses on the wide seafloor. In this way, the Antarctic Bottom Water has created good oxidation conditions and production places for the growth of polymetallic nodules and Co—rich manganese crusts, and also has supplied a large quantity of nodule—forming materials.

CONCLUSIONS

From the above analysis of the Pacific paleooceanographic events and mineralizations of ferromanganese oxides, we come to the following conclusions:

(1) Lithospheric plate movement changed distribution pattern of ocean and continent, influenced and controlled evolutions of global climate and oceanic current systems.

(2) Polymetallic nodules and Co—rich manganese crusts are products of " multiple event deposition" in given settings, that is, in certain geologic periods when global climate became cold, the Antarctic Bottom Water was active, bioproductivity increased and hiatus took place, and thus we can say that the formation and distribution of the nodules and crusts depend upon marine environment and geologic events.

References

1. P.F., Barker, J., Burrel, The opening of Drake Passage. Marine Geology. Vol.25, P15—34.(1977)
2. R.H., Benson, Miocene deep sea ostracoda of Iberian Portal and the Balearic Basin. Marine Micropal. 1, P249—262.(1976)
3. N.A., Brewster, Cenozoic biogenic silica sedimentation in the Antarctic Ocean, based on two deep sea drilling project sites. Geol. Soc. Am. Bull. Vol.91, P337—347.(1980)
4. A.N., Carter, Phosphatic nodules beds in Victoria and the Late Miocene—Pliocene eustatic event. Nature. No.276, P258—259.(1978)
5. M.B., Cita, Early Pliocene paleoenvironment after the Messinian salinity crisis. VI African Micropaleontolog Colloquim, Tunis 1974.(1976)
6. D.S., Cronan, Unterwater Minerals. Academic Press, London. P1—363.(1980)
7. P., Halbach, Processes controlling the heavy metal distribution in Pacific ferromanganese nodules and crusts. Geol. Rundsch. Vol.75, P235—247.(1986)
8. P., Halbach, U., Hebisch, Ch., Scherhag, Geochemical variations of ferromanganese nodules and crusts from different province of the Pacific Ocean and their genetic control. Chem. Geol. Vol.26,No.1,P3—17.(1981)
9. P., Halbach, C., Scherhag, U., Hebisch, V., Marchig, Geochemical and mineralogical control of different genetic type of deep sea nodules from the Pacific Ocean. Mineral Deposita (berl). Vol.16, No.1,P59—84.(1981)
10. P., Halbach, D., Puteanus, The influence of the carbonate dissolution rate on the growth and com-

position of Co—rich ferromanganese crusts from Central Pacific seamount areas. Earth and Planetary Science Letters. vol.68, P73—87.(1984)

11. P., Halbach, M., Segl, D., Puteanus, A., Mangini, Relationship between Co—fluxes and growth rate in ferromanganese deposits from Central Pacific seamount areas. Nature. No.304, P716—719.(1983)

12. B.U., Haq, Biogeographic history of Miocene calcareous nannoplankton and paleooceanography of the Atlantic Ocean. Micropaleontology. Vol.26, P414—443.(1980)

13. J.R., Hein, W.A., Bohrson, and M.S., Schulz, Variations in the fine—scale composition of a Central Pacific ferromanganese crusts: paleooceanographic implications. Paleooceanography. Vol.7, No.1, P63—77.(1992)

14. K.J. Hsu, et al., Mediterranean salinity crisis. In Initial Reports of the DSDP. VOl.42, P1053—1078.(1980)

15. K., Ikari, Nishimura, Geologic history of the Tenpo seamount of the Nishi—Shichito Ridge, the Izu—Bonin Arc. Bull. Geol. Surv. Jpn. Vol.42,P19—41.(1991)

16. L.D.Jr., Keigwin, Pliocene colsing of the isthmus of Panama, based on biostratigraphic evidence from nearly Pacific Ocean and Caribbean Sea cores. Geology. No.6, P630—634.(1978)

17. G., Keller, Middle to Late Miocene planktonic foraminiferal datum levels and paleooceanography of the north and southeastern Pacific Ocean. Marine Micropaleontology. Vol.5, P249—281.(1980)

18. G., Keller, Miocene biochronology and paleooceanography of the north Pacific. Marine Micropaleontology. Vol 6, P535—551.(1981)

19. G., Keller, J.A., Barron, Paleooceanographic implications of Miocene deep sea hiatuses. Bull. Geol. Soc. Am. Vol. 94, P590—613.(1983)

20. G., Keller, JA., Barron, Paleodepth distribution of Neogene deep—sea hiatuses. Paleooceanography. P697—713.(1987)

21. F.M., Kemp, Palynology of Leg 28 Drill Sites, Deep Sea Drilling Project. In Initial Reports of the DSDP. Vol.28,P599—623.(1975)

22. J.R., Kennett, Cenozoic evolution of Antarctic glaciation, the circum—Antarctic Ocean and their impact on global paleooceanography. Journal of Geophysical Research. Vol.82, No.77, P3843—3859.(1977)

23. J.P., Kennett, P., Vella, Cenozoic planktonic foraminifera and paleooceanography at DSDP Site 284 in the cool subtropical South Pacific. Initial Reports of the DSDP. Vol.29, P769—782.(1975)

24. J.P., Kennett, D., Watkinsr, Regional deep—sea dynamic processes recorded by Late Cenozoic sediments of Southeastern Indian Ocean. Geol. Soc. Am. Bull. Vol.87, P321—339.(1976)

25. T.C. Jr., Moore, T.H., van Andel, C., Sancetta, N., Pisias, Cenozoic hiatuses in pelagic sediments. micropaleontology. Vol.24, P113—138.(1978)

26. M., Leinen, Biogenic silica accumulation in the central equatorial Pacific and its implications for Cenozoic oceanography. Geol. Soc. Am. Bull. Vol.24, P1—22.(1979)

27. S.M., Savin, The history of the earth's surface temperature during the last 100 million years. Ann. Rev. Earth Planet. Sci. Vol.5, P319—355.(1977)

28. S.M., Savin, R.G., Douglas, and F.G., Stehil, Tertiary marine paleotemperature. Geol. Soc. Am. Bull. Vol.86, p1499—1510.(1975)

29. N.J., Shackleton, J.P., Kennett, Paleotemperature history of the Cenozoic and the initiation of Antarctic glaciation: oxygen and carbon isotope analyses in DSDP Sites 277, 279 and 281. InitialReports of DSDP, Leg 29. P743—755.(1975)

30. A., Usui, and Takashi, Fossil manganese deposits buried within DSDPODP. Core. Legs 1—126, Marine Geology, 119, 111—136(1994)

31. P.R., Vail, J., Hardenbol, Sea—level changes during the Tertiary. Oceanus. Vol.22, P71—79.(1979)

32. T.H., Van Andel, G.R., Heath, G.R., and T.G., Moore, Cenozoic history and paleooceanography of the central Equatorial Pacific Ocean. Geol. Soc. of Amer. Memoir. Vol.143, P1—134.(1975)

33. J.T., Van Gorsel, S.R., Troelstra, Late Neogene planktonic foraminiferal biostratigraphy and climostratigraphy of the Solo River Section (Java, Indonesia). Marine Micropaleontol. Vol.6, P183—209.(1987)

34. J.K., Weissel, D.E., Hayes, and F.M., Herron, Plate tectonics, synthesis: the displacement between

Australi, New Zealand and Antarctica since the Late Cretaceous. Marine Geology. Vol.25, P231—277.(1977)

35. S.I., Andreev, Law of formation of ferromanganese concretion on the world Ocean. Geology and hard minerals on the world ocean L. NiiGA. 30—40.(1980)

36. M.A. Levitan, and A.P. Lisitsion., distribution of ash layer in loose sediments on the Pacific Ocean. Dokl. Sci. USSR. V.241. No.4.(1978)

37. N.C. Skornakova, Ocean ferromanganese concretion (law of distribution and composition). Disk D.G.—M.N.M 68

38. J.A., Wolfe, An interpretation of Tertiary climates in the Northern Hemisphere. Am. Sci. Vol.66, P694—703.(1978)

39. Xu Dongyu, Geochemical characteristics of manganese crusts in the Central Pacific Ocean. Marine Geology & Quaternary Geology. Vol.10, No.4, P1—10.(1990) (in Chinese)

40. Xu Dongyu, Yao De, and Chen Zongtuan, Paleooceanographic environments and events of the formation of manganese nodules. in Resource Geology Special Issue. No.17, P66—75.(1993)

41. Xu Dongyu, Geochemistry and Genesis of Polymetallic Nodules. China Geological Publishing House, Beijing. P1—123.(1993) (in Chinese)

42. Xu Dongyu, Chen Zongtuan, and Meng Xiangying, Late Cenzoic paleooceanographic environment and events of the Central Pacific Ocean. China Geological Publishing House, Beijing. P1—176. (1994) (in Chinese).

43. Xu Dongyu, Jin Qinghuan and Liang Dehua, Polymetallic nodules and its formation environment in the Central Pacific Ocean. China Geological Publishing House, Beijing. P1—418.(1994) (in Chinese).

44. Xu Dongyu, Yao De, Liang Hongfeng and Zhang Lijie, Paleooceanic environment for formation of polymetallic nodules. China Geological Publishing House. P1—111.(1994) (in Chinese).

45. D.A., Dunn, and T.C. Jr., Moore, Late Miocene / Pliocene (Magnetic Epach 9—Gilbert Magnetic Epoch) Calcium—Carbonate stratigraphy of the equatorial Pacific Ocean. Geol. Soc. Am. Bull. V.84 2021—2034.(1981)